The Practical Inventor

J.T. Wilkinson

Illustrated by:

Shannon Parish
Joe Wilkinson
J.T. Wilkinson

Veritek Publishing, LLC
Littleton, CO

Published by:
Veritek Publishing, LLC
PO Box 630253
Littleton, CO 80163
www.veritek.com

The Practical Inventor
©2007 by JT Wilkinson
Printed in the United States of America

*"You can't just sit on an idea and expect to keep ownership of it. The patent goes to the person who can show that they were first to invent and **were diligent in pursuing a patent** within a reasonable amount of time (usually a year or two)."*

       Brian Kunzler
       Patent Attorney

"An inventor is someone who asks why of the universe and lets nothing stand in the way of the answers."

       John Galt

ISBN: 978-1-4565336-5-6

Liability Disclosure: Although every effort has been made to provide accurate information, none of the information contained within should be construed as legal advice in any form. Advice from an intellectual property attorney should be obtained and is highly encouraged.

No part of this publication may be reproduced, stored in a retrieval system, or transmitted in any form or by any means, electronic, mechanical, photocopying, or otherwise, without the prior written permission of Entrepreneur Resource, LLC. The material in this book is furnished for the informational use only and is subject to change without notice. Entrepreneur Resource, LLC assumes no responsibility for any errors or inaccuracies that may appear in this book.

Acknowledgments

Standing up from the circle and looking apprehensively at the other faces looking at me I start, "Hello, my name is Jim Wilkinson and I am an obsessive-compulsive inventor." Wow, it has taken me a long time to admit that in public! OK, I am an inventor and like a lot of other inventors, not only believe that I can but actually do just about anything I set my mind to. However, it has taken me over 50 years to learn and embrace the fact that there are others out in the world who might be able to do parts of what needs to get done much better or at least faster than I could. None of us are good at everything and for those who say they are, enough said.

There are a lot of people who have inspired, helped, encouraged, and supported the efforts to put this book together. Many of these people share the same desire to help the individual inventor move their ideas into the product phase where the inventor has a profitable product.

Thanks go to Berny Dohrmann of CEO Space (www.ceospace.biz). CEO Space is a great organization that teaches CEO's how to better operate their businesses, raise capital, and provides an extraordinary platform for networking. They hold a 10 day conference 5 times a year that is equivalent to an MBA. It was at one of these conferences I decided that this book and kit are needed.

My friend John Quiring who has stood behind my crazy adventures for over twenty years, all I can say is thank you. Without his support, I would have never completed this.

All my other friends, too many to list you all but you know who you are. Without your support, proof reading, content review, none of this would have come together.

Thank you all.

Getting the most from this book.

This is not the kind of book that you just sit down and read. This as a roll up your sleeves and get your hands dirty, step-by-step kind of book. My greatest compliment will be to see your book battered, dog eared, coffee stained, maybe pocked here and there with the stray drill bit. OK, enough already, the title of this section is something about how to get the most from the hard earned dollars (US) that you spent to acquire this book and you want to know how in the world you are going to make use of it.

1. Familiarize yourself with the book.
2. Make a pot of coffee, get a snack, get comfortable and read through the book without doing any of the exercises. At the very least, skim through it so you can see what is inside. You may want to just dive in but by taking this step you will have a better picture of what's going on as you do the exercises.
3. Roll up your sleeves and start with the Introduction.
4. Just do it! I can guarantee that after you do an exercise and revisit it 6 times, it is still not going to be perfect, at some point it is good enough so get on with it already.
5. Do your best to complete each exercise. There will be parts you cannot complete but once you've built your team, come back and complete it with your team.
6. Stay in sequence. Do Chapter 3 after 2 after 1 and so on.
7. If you get completely stuck, send me an email through The Practical Inventor website.
8. Once you've gone through the book and completed each exercise, keep reviewing them and revising your responses, as needed to move your project forward.
9. Use the resources and contacts in Appendix A to help move your project forward.
10. Use this book as a reference book and come back to it often.
11. Use the notes pages to keep notes but be sure to transfer them to your design journal.

This book is dedicated to all those inventors who refuse to quit and refuse to listen to the naysayers.

Contents

Introduction 1

Chapter 1 – Documentation 13

Chapter 2 – Ideation 23

Chapter 3 – Feasibility 39

Chapter 4 – Intellectual Property 49

Chapter 5 – Plan 79

Chapter 6 – Prototype 95

Chapter 7 – Launch 137

Appendices 151

Appendix A – Resources 153

Appendix B – Inventing the Inventor 155

Appendix C – Business Overview 167

Introduction

Introduction   The
                                               Practical Inventor

Introduction

Welcome to the Practical Inventor. This is your guide to take your idea from the drawing board to a salable product. But before we start that journey, let's consider some ways in which this manual will benefit you as a practical inventor.

Before starting, let's talk a bit about this book, you may want to consider purchasing the complete kit at www.thepracticalinventor.com. The full kit contains a more extensive manual, a CD with electronic versions of the exercises and templates, a Practical Inventor Design Journal for creating proper documentation of your project, and Practical Inventor workbook to provide additional guidance to organize the various documents and literature collected during your quest.

Sad to say, many inventors fail to make money with their inventions. Even when an invention is a fantastic step forward in technology or solves a serious problem it may not live up to its potential because the market isn't quite ready for it. More often, however, this failure occurs because the inventor lacks the business savvy to effectively market it. Designing and building a great invention is quite different from turning that invention into a commercial success. This manual and the balance of the Practical

The Practical Inventor

Introduction

Inventor kit provide a map to take you down that road from idea to product to profit.

I've designed the text and exercises in this manual to help the individual inventor (or small business) move from idea to revenue-generating product with the minimum amount of time and resources. My thirty years' experience working with fortune 100 companies has provided the skills and knowledge contained in these pages. Adapting that experience to the needs of the individual inventor was simple because the success process is very similar. The primary difference being that the individual inventor lacks the corporate structure to support this development process.

As you work your way through this manual you will likely find yourself skipping portions of an exercise and come back to it after completing another chapter or exercise. Do not be alarmed when this occurs. The exercises are designed to be ongoing. In order to get the most benefit, use the files contained on your CD and save the files to your hard drive in the appropriate folders. You should be returning to these files regularly as you learn new things. Print out a copy of each exercise as you add to it and insert it into your workbook in the appropriate sections. This will also enable you to better collaborate with your team.

The Practical Inventor manual is the definitive guide for those interested in starting or growing their businesses based on their inventions. Even if you are thinking of licensing your invention, your value will be maximized by creating a company and building value following the SIMPLE™ process. We will focus on product development, synchronizing it with the business development. Realistically, costs are incurred at different stages, but careful planning and well-considered strategies can reduce the financial impact of these times. (Of course, we help you accomplish this.) The ultimate goal is developing your idea into a viable product (or service) that can be sold to customers or licensed to a third party for the purpose of profit.

Let's touch on patents for just a moment. It's important at this point to know whether you want a patent for the sake of having a patent or whether you want to profit from your idea. Sometimes it is to your advantage not to get a patent and create the public disclosure of key proprietary technology. I probably have 6 profitable ideas for every one patent. For various reasons, I choose not to pursue a patent, yet I've made good money off several of

Introduction
The Practical Inventor

these ideas. Very likely you use some aspect of one of my products every day.

In Chapter 4 we will also learn about something called Intellectual Property (IP). It's vital that you know what it is as well as different methods of protecting it rather than focusing strictly on the patents.

By now you may be wondering if you will make overnight fortune. That does happen but is also very rare. I have yet to figure out how to do it. If I ever do, rest assured, the next revision to this manual will be on the shelves in short order.

My business mentor of many years used to remind me "money is nothing more than a tool, a very important tool but just a tool." Considering that perspective, we may find bigger things in life that we should be aspiring to, such as focusing on the money as a facilitator to get us there.

Speaking of money, product development and prototypes can become significant money pits. The SIMPLE™ process presented in this manual will demonstrate a strategic overview for developing products in an effective and expeditious fashion while staying within a reasonable budget.

Sometimes it is to your advantage not to get a patent

Included is a brief discussion on different prototype technologies, their practical applications, pros and cons, and relative costs.

In the early 1980's, I was a partner in a small marine electronics company where I was the midnight elf who came in at night and repaired customer products after working my day job. My partner and I joked about the boats being nothing more than a hole in the water into which people poured money. This could result from the boat owner's having more money than sense, or from knowing exactly what they wanted despite what they wanted contrary to expert advice. Almost always it was because the boat owner did not have a well thought-out and written plan, operating instead by what seemed to look good at the time. Isn't it ironic that the very same holds true with an inventor? We will see the value of a carefully considered plan so we don't get sucked into that bottomless hole in the water.

The Practical Inventor

Introduction

Have you ever played the game where several people get into a line and someone at one-end whispers a phrase into the first person's ear? What the last person hears often bears no resemblance to the initial phrase. How does this apply in our context? Have you ever tried to explain your idea and the person you are talking to just doesn't get it? You must be able to define your idea clearly, specifically, and in definitive terms that don't "get lost in the translation."

Exercise 0-1: Define your idea

Welcome to your first exercise. Let's keep it easy to cut your teeth. Answer the following questions as fully and completely as you can. Use sketches and pictures if you like.

What problem in society does your idea address and solve?
 How does your idea solve this societal problem?
What products result from your idea?

Introduction The Practical Inventor

A tale of two inventors

The boy and his mother sat in the Principal's office. The Principal explained to the mother "You need to do your son a favor and put him into a vocational school and teach him a trade so that he has a chance of making it through this world because he will never make it here."

This young man, Thomas Edison by name, traveled through life with his eighth-grade education. He went on to invent such products as the light bulb, phonograph, and practical rechargeable batteries. At some point in the operation of his company, he hired a Lithuanian immigrant as a lab assistant. This fellow was Nikola Tesla (the father of radio). Tesla worked for Edison for a few years but like most inventors I know today, had his own way of doing things and eventually left the confines of Edison's lab. Later on, Tesla and Edison got into a heated competition for the electrification of the United States. Tesla wanted Alternating Current (AC) electricity for power distribution and Edison wanting Direct Current (DC) electricity. Tesla won with the contract award from Westinghouse to design a hydroelectric power plant at Niagara Falls.

What's the point here? We can learn a lot from each of these great men who typify the two basic types of inventors. Edison almost boastful of taking 10,000 attempts to arrive at a working light bulb was not afraid to act on his idea, learn from his mistakes, and try again. Tesla, on the other hand, scoffed at such an excessive approach, saying that he would have thought about it for a bit and then just done it. Not afraid of acting on his ideas, he was much more intellectual and calculating than his former employer. Unfortunately, he was also a man far ahead of his time. Tesla died almost penniless after envisioning technology that only today are we able to see come to fruition (turbine engines and particle beam weapons

So what is an inventor? An inventor is someone who has an idea and does something about it. It may be someone working for a large corporation, or an individual who is 7 or 70. Inventors possess varying degrees of ability in areas such as building

> Many inventions are the result of solving a problem.

prototypes, engineering, business development, or finance. Rarely, however, do these skills include adequate expertise in all areas needed. In fact, the demise of many inventions is the failure of the inventor to let go,

The Practical Inventor   Introduction

get out of the way, and allow others to make the invention, and often dream of the inventor, into a successful product.

In Chapter 7 we will discuss launching your product. We will address whether to license or take your product to market directly. Numerous factors will be considered but suffice to say that yes, your idea may be worth billions of dollars someday, BUT NOT TODAY. The value will come in the ability to move your invention successfully to the marketplace in a form that people (or companies) are willing to buy.

Don't sell yourself short because you are not an engineer or that you don't have this or that training. I spent almost half of my professional career without a formal degree and finally finished a Bachelors of Science in Industrial Technology, not even engineering. Take stock of what you can do and what you love doing, then just get busy. Those who have ever worked for me would tell you that I'd rather have them doing something and making mistakes than stuck analyzing and redesigning to avoid mistakes. (Does this make me an Edison type?) Just learn from the mistakes and try to not spend too much time or money in the process (shades of Tesla here?). This manual is designed to help you avoid the typical errors.

Another vital consideration is embodied in this question: what is good enough? Your product does not need to be perfect to launch. How many new cars have you seen that had absolutely perfect designs and are built perfectly? If your product is perfect, where will you have room for later enhanced products to launch and sell to your now repeat customers?

One quick note on self-image: An inventor friend of mine has a hang up about her weight and "ugly" legs. Yet she is a very attractive woman and her legs are just fine. While she will likely never be a runway model, she has nothing to be self-conscious about. Many of us have experienced deep-

Introduction   The
                                      Practical Inventor

core image issues. If we define these issues, embrace them, celebrate them, we can become unstoppable! Buried and repressed, these issues affect body language and mannerisms that are picked up and misinterpreted by others. If they think we are hiding something, this will likely impede our ability to make the alliances necessary to move our idea forward. Embrace and celebrate the uniqueness and individuality, then look at how these can be used to our advantage. You will discover that you have taken your ability to be creative and think freely to a new level. (My own story on such issues is the subject of my next book.)

So you have an idea and just don't know where to start? It doesn't matter whether you are a lone inventor or an employee of a large corporation, getting started properly with your idea is often a challenge. It all comes down to having an effective and efficient process to follow. I call it a Product Development Process and guess what? A lot of large corporations don't have a plan either.

The diagram below presents the general process of inventing and developing a product. Each stage is introduced here and discussed in detail later in other chapters.

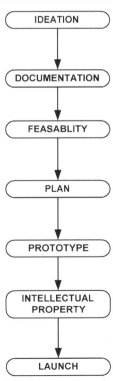

Ideation

So what is ideation? It is the process of defining your idea. Where do ideas come from? Many arise from need. For example, you are on a camping trip and have a problem getting the pot off the fire. Suddenly you are saying to yourself, "If I only had something to hold this pot above the fire, it would really make cooking easier." Congratulations, you are an idea person. If you want to be an inventor, find some poles, tie them together, and hang your pot above the fire. Now you have invented something and built a proof of concept prototype.

Inventions come from many sources. The most common involves seeing a problem and envisioning a product or improvement to an existing product. But where do you start? We will work on that together, one step at a time. Here's where ideation comes into play. First, properly document the idea, and then perform an initial assessment of the idea to determine if the idea is worthwhile moving forward with before you spend time and money. Some friends of mine came up with the concept of "fail

The Practical Inventor

Introduction

fast". The idea here is to eliminate ideas and projects quickly rather than burn time and money without producing positive results. You are probably sitting there saying "well duh!" but many of us hold onto a project, pushing it forward because we believe it is a great idea. Unfortunately, the reality is that someone already did it or it's not practical. Letting go is not easy so we are going to keep coming back to "fail fast" throughout the process.

Documentation

Proper documentation is the first step to protecting your idea.

The design journal is the tool for documenting your idea and the journey it takes from inception through launch. The answers to the following questions form the initial basis for product specification, product feasibility, market potential, and patentability. Each of these will be discussed in more detail in later Chapters.

1. Define the product or idea in one or two paragraphs. Draw a couple of sketches if applicable. If you have napkin sketches, now would be the time to put them into your journal. Document who you talked with about your idea and when

2. What was your inspiration for this idea?

3. Who would use it and for what?

4. What would potential customers expect from your product? We inventors tend to say "well of course they would want this!!!" We will do a reality check.

5. How much might they be willing to pay for it? What is your basis for this amount?

6. Are there similar products on the market? Include photos.

7. What kind of intellectual property (IP) protection is appropriate for your idea? Start thinking about whether or not you need a patent? We will address patents and IP protection in chapter 4. The important thing now is that it is not the time to run out and spend money on a patent search.

This all may sound a bit overwhelming. Don't worry; it looks a lot worse than it is. As you progress through our SIMPLE™ process it will make more sense. Hang in there and it will come together down the road. Don't worry if you cannot answer all of the questions right now. Even if you did answer all of the questions right now by the time we finish, you will either change what you write or discover how incomplete your answers are. That is OK. Put as much detail into it right now as you are able, and go from there. That's why it is called development.

The Plan

Did you ever see the movie *The Hunt for Red October*? There was this great line in it, "The Ruskies don't take a dump without a plan." I love that line and it is pretty good insight. A good plan is essential for moving an idea forward in an expeditious manner and without spending a fortune. Your initial plan needs to be one of the first entries in your design journal on this idea. This plan should address the answers from the documentation stage.

Think of it this way; you have a great idea that you think would revolutionize the world and make you billions of dollars. Unless your plan takes into account the target audience along with their preferences and desires, the product is not going to sell. Potential has no inherent value in a product, sales and profits say it all.

Having mentioned sales and profits, this seems a good time to talk about not spending money. Most inventors begin their inventing lives with little or no cash on hand. Companies too, have limited resources, so it makes sense to address this in a structured and methodical manner. I call this a product development plan. The plan I will share with you has been developed through thirty years experience with many different products and industries.

The Practical Inventor

Introduction

Feasibility

Think about the concept of "failing fast". Some associates have told me not to use this name because of the negative connotations. Consider this: If your idea was already done in 1960 and failed, wouldn't you want to know about it? It does not mean drop the idea and move on but it does mean you may not get a patent on it. There may be some new, enabling technology or cultural timing that would make this idea successful today.

The concept around failing fast is just like selling. The mark of a good salesman is the ability to get quickly to the "no". Why spend a lot of time and resources on something that is not going to bring a return. Before spending time and money on prototypes, patent searches, and the like, find out if your idea is something that is practical and can be profitable. If not, drop it and move on. For every good idea, I have probably had 5 or 6 that I've left behind. One time, I had an idea for a product only to find out that it was in development at NASA. How cool is that? When that happens to you, don't be discouraged; in fact, be encouraged because you are on the right track and in great company.

Be realistic. Don't try to rationalize that an idea is good just because it's yours. It may not be feasible. Look at it critically and from your potential customer's perspective. Ask yourself, why would you spend hard earned dollars for this? Do not assume you will sell thousands of them just because there's nothing like it on the market. You are also competing with pizza, shoes for the kids, even a tank of gas.

Prototype

We are going to have some fun in chapter 6!

Prototypes require a bit of planning before the build begins. I previously mentioned my work at the small marine electronics company and the joke about a boat as a hole in the water to which the owner pours money. The not so funny part is that I have seen inventor after inventor that you could say that about their idea! Don't be one of them!

A fortune need not be spent on building a prototype; whatever the reason for building it. If you just want to see if the idea can work (Proof of Concept), it can be made from duct tape and cardboard. A friend of mine just showed me a prototype of a pet related invention of hers. She made if from about 14 inches of duct tape and items salvaged from the trash can. It even looks pretty good. If she wants to exhibit this at a trade show however, she would consider one made from a more expensive Rapid Prototype process, painted, and with some nice decals applied to mimic a production product. Cost is a consideration here. Her proof of concept cost her about 30 minutes of her time and 14 inches of duct tape, less than one dollar in materials.. The trades show model will cost between $900-1100 by the time it is finished painted, and decals applied. A production unit will probably be $3-6 and will require $15,000 to $18,000 in tooling. You can now see the reason for knowing your purpose and audience for this prototype?

Intellectual Property

A lot of money can go out very quickly when you start engaging professional help protecting your intellectual property. Many inventors rush into this expense because they believe that if they could just get a patent on their idea, they can make a fortune. Unfortunately, this very rarely happens,. The patent is a very overrated item but can be very valuable when appropriate. The big questions are how to "protect" your IP and when. For now, do not run out and hire someone to perform a patent search. Please read chapter 4 on IP before making any decisions.

The Practical Inventor

Introduction

Launch

A good friend of mine would call this the "phases and gates" part. We will look at the product development process as the main tool in moving your idea from a prototype to a product ready for delivery to the customer. We will address a number of pertinent questions including:
- What are the pitfalls?
- What is the scalability?
- What type of manufacturing process should be used?

At this juncture, it's essential that we answer the right questions if we want to succeed. However, as Ambassador Sarek said in Star Trek IV, "It is hard to have an answer when one does not know the question."

You are going to have lots of homework in this chapter identifying the questions and the answers for your particular situation. Fortunately if you have done the work in the earlier chapters outlined in this manual, you have already collected much of the information needed for decisions during the launch phase of your idea.

That's it! Now it is time to quit talking about your idea and get to it. Roll up your sleeves, warm up your coffee and let's go.

Chapter 1

Documentation

Documentation   The Practical Inventor

Documentation

Regardless of the strategy you choose for protecting your idea, it still needs to be documented in a proper fashion.

Poor Man's Patent

You have most likely heard about the "poor man's patent"? You write up your idea, seal it in an envelope, and mail it to yourself. Of course you don't open it after it arrives. One inventor I worked with many years ago taught me this method. He even had a special way of folding the paper. Then his patent attorney who was a patent examiner for many years, and also one of the primary patent attorneys for General Motors told me not to waste my time and proceeded to teach me his method for documentation. Since then (about 30 years ago now) I've had several inventors and a few patent attorneys tell me about the poor man's patent but most patent attorneys I've talked to share the opinion that it is a worthless method. Regardless whether you see the poor man' patent as viable or not, it is not going to hurt anything by doing it. I would just suggest not relying on it for the documentation of your idea.

Computers and Electronic media

The best method for documenting your idea is still the bound journal. Even in the electronic age, the written journal reigns.

Don't get me wrong, I use and rely on my computers almost every waking moment but it just isn't the proper tool for this job. If you know computers well enough, you know that you can change the calendar date on the computer to any date you like and back date data very easily. Because of the ease of this process, computer documentation alone might be questioned in some circles were an issue to arise.

> Proper documentation is the first step to protect your idea.

My journal is filled with insertions from documents and files I've printed out, and still have the original on the computer. This is my suggestion for using computer data, even if it means your journal entries end up being nothing but printed out log entries from your Outlook journal, you are still creating an irrefutable document that has verifiable corresponding data on the computer.

The Journal

Unlike the poor man's patent, a design journal is irrefutable and invaluable when, at some point you decide to defend your invention when someone copies it (IP infringement). This journal should be a book such as an accounting ledger book with the pages sewn in, not glued, and each page numbered. Books with lined pages that meet these specifications are relatively inexpensive and easy to find. I've used basic accounting ledger books from the drug store. If you want graph paper pages, a university bookstore is usually the best source. My preference of course, is The Practical Inventor Design Journal.

Documentation   The
                                          Practical Inventor

Some rules for using and handling your journal:

1. This journal is going to contain your most prized and closely guarded secrets! Do not let it out of your possession to anyone at any time!

2. Document everything about your idea:
   - Photos
   - E-mails
   - Notes
   - Research findings
   - Executed non-disclosure agreements (NDA)
   - People you talk to
   - Schematics
   - Sketches

3. Use your journal to tell the story about the development of your idea from start to end.

4. If you are working on more than one project, use the same journal for them all, not a separate journal for each.

5. Oh, did I say to guard your journal with your life?

6. Don't worry about keeping the journal pristine. You want it to have the appearance of normal wear and tear. Don't abuse it but an occasional coffee stain and solder iron burn is good for the soul.

7. Never remove any pages from your journal.

Journal General Entries

On the outside front cover, number and date your journal so the cover identifies your journal as #1 or #2 or so on and include the start date for the journal. With The Practical Inventor Design Journal, there is a place on the spine of the journal provided for this. When you fill this journal, add the ending date. Keeping an index in the rear of the journal by product title is invaluable when wanting to refer back to information at a later time. This isn't needed for actual documentation, just sanity later on.

On the inside of the front cover, there should be some ownership identification. I use:

This book is the property of

JTW, LLC.

and is being used by J.T. Wilkinson
for log entries of:

1. Research data
2. Business activity specific to a project
3. New ideas and concepts
4. Progress Reports

Each entry shall be dated and signed.
Each page shall be witnessed and dated.

New ideas and concepts should be formatted:

    Title
    Purpose
    Description
    Drawings
    Ramifications
    Test data and development information
    Photos
    Novel Features

The information contained herein should be sufficient that someone competent in the field can recreate the work.

You can write something like this out by hand or create it on your computer and paste it in. Either works fine.

Documentation	The Practical Inventor

Exercise 1-1:
Gentlemen, Start Your Journals:

Obtain a journal. If you do not have a Practical Inventor Design Journal, be sure that the journal you use has the following characteristics:
   a. Pages sewn into the binding.
   b. Each page numbered sequentially.

If you have a Practical Inventor Design Journal, fill out the information on the inside front cover.

Creating Documentation

Each invention needs the following items in the documentation. However, all this information need not be on the same page. The two most important features of your journal are that it documents your idea and presents a story about the development of your invention from that first idea. A journal that does both of these is the most valuable.

- **Title** – The name of your invention
- **Purpose** – What society issue you are addressing and what the invention is intended to do
- **Description** – Full description with sketches, schematics, diagrams, photos that fully describes not only the invention, but also completely details all the work leading up to and progressing through the development.
- **Ramifications** – What it would mean to someone using it.
- **Possible novel Features** – Describe what is unique and original about the invention.
- **Advantages** – What are the advantages of this invention compared to similar products already available?

Each page must be **signed, dated, and witnessed** – usually by two people. The witnesses need to be able to understand what they are witnessing. That means that someone who knows nothing about electronics is probably not a good witness for pages dealing with the technical aspect of an electronic device.

> The witness must be able to understand what they are witnessing.

The Practical Inventor
Documentation

A topic or title almost never fits onto one page. Simply continue a topic on following pages with the original title and date on each following page. Remember, each page must be signed and witnessed. You can also save yourself a lot of time by starting your index in the back of the journal right away.

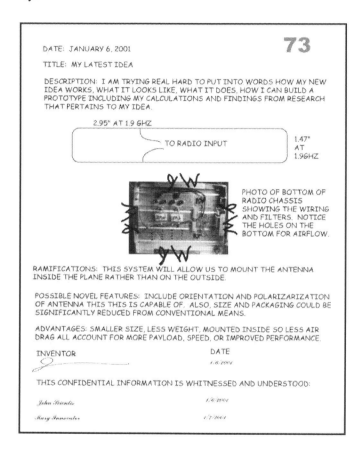

This is an example of a page showing some opening thoughts, a sketch, and a photo pasted into the page. Notice that the pasted in photo is initialed all the way around. The initials make it very difficult to tamper with the pasted in photo. This works for any other type of pasted in item such as a napkin sketch or email.

Documentation   The
                                        Practical Inventor

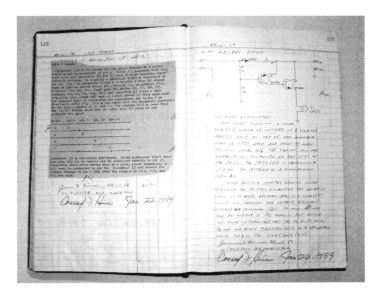

The above photo is taken from one of my old journals. Notice how the pasted in pages are initialed and the pages are signed and dated. This journal was an old (really old) ledger book that my accountant gave me because it was too old to be of use to him.

Don't worry about having the pages numbered when you buy a journal. That would be ideal but you can also just go through the whole journal when you set it up and number the pages yourself.

The Practical Inventor   Documentation

Exercise 1-2: Document your idea:

1. Write the following into your journal:
   - **Title** – The name of your invention
   - **Purpose** – What society issue you are addressing and what the invention is intended to do
   - **Description** – Full description with sketches, schematics, diagrams, photos that fully describes not only the invention, but also completely details all the work leading up to and progressing through the development.
     - ❖ Define the product or idea in one or two paragraphs. Draw a couple of sketches if applicable. If you have napkin sketches, now would be the time to put them into your journal. Document who you talked with about your idea and when
     - ❖ What was your inspiration for this idea?
     - ❖ Who would use it and for what?
     - ❖ What would potential customers expect from your product? We inventors tend to say "well of course they would want this!!!" We will do a reality check.
     - ❖ How much might they be willing to pay for it? What is your basis for this amount?
     - ❖ Are there similar products on the market? Include photos.
     - ❖ What kind of intellectual property (IP) protection is appropriate for your idea? Start thinking about whether or not you need a patent? We will address patents and IP protection in chapter 4. The important thing now is that it is not the time to run out and spend money on a patent search.
   - **Ramifications** – What it would mean to someone using it.
   - **Possible novel Features** – Describe what is unique and original about the invention.
   - **Advantages** – What are the advantages of this invention compared to similar products already available?

2. What other ideas do you have? Get them written into your journal.

3. Sign and date each page. Have each page witnessed by someone who will understand what you have written.

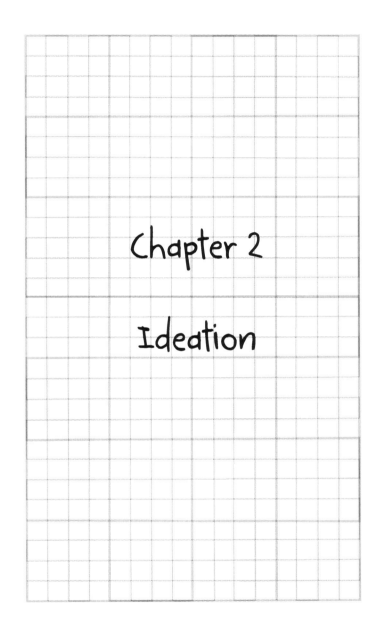

Ideation

Ideation

Ideation is the process of defining your idea and organizing it into a workable state. The ideation process was created to help you start your journey down the road toward production and profit.

The Ideation Process

The following sections of this chapter discuss in detail the Ideation process and walks through exercises to accomplish each of the steps.

Select the Project:

Many inventors have a lot of ideas but can't decide on which idea to work on first. As a result, they tend to bounce from one to another, rarely taking any of them to the point of being ready for the market as a product. One of the most necessary but also most difficult requirements for most of us becomes choosing one project and focus on it alone.

The following exercise is the project selection matrix. In this exercise you will list and rate your ideas to help pick the one idea that can produce positive cash flow in minimum time and monetary investment. Think also about the bigger picture. Let's suppose you have a lot of ideas but haven't had any take off yet. What if you were to sacrifice (so to speak) one of your ideas by raising investment capital on it and getting that project to the point of generating good revenue? Then you could have the resources to self fund your next idea and so on (not to mention all the false gods who get appeased by the sacrifices).

As you learn and grow, it is likely that you will discover the advantages of sharing your opportunities with others. By doing this, you will find it much faster and easier to launch your future products by leveraging others capital rather than digging into your pockets. I mention this to emphasize the fact that all of the processes we talk about are valid whether you are using your capital or investor capital.

Ideation

The Practical Inventor

Exercise 2-1: Project Selection Matrix:

The following matrix is designed to help select the one product that can produce positive cash flow in minimum time and monetary investment. List each of your ideas in the IDEA column and score each idea across the row beside it. Don't worry about getting the perfect score because you always can and will change it later, just get started. Also, there are no perfect answers, only your best guess at the time. Remember that the difference between you and any ten other "inventors" is that you are taking action on your ideas. Of course, if you are the one in a million inventors that has someone to fund the development of a particular product, just ignore this section and work on that product for now..

PRODUCT SELECTION MATRIX											
IDEA	DESIGN	PROTOTYPE	DEVELOPMENT COST	TOOLING	INTELLECTUAL PROPERTY	CERTIFICATIONS	UNIQUENESS	LAUNCH COST	MARKET READINESS	PROFITABILITY	SCORE

Score

Total up the numbers in each row. The highest score is likely to be your best candidate to focus on. Just remember that this is a guide, not a chiseled in granite rule. Also remember that it will almost always take you longer and cost more than your best guesses.

The Practical Inventor   Ideation

Perform Initial Research and Assess Feasibility:

At this point in the development of your idea, the initial research and assessing feasibility are so closely aligned that we will cover them in the same section. As the term implies, the initial research generates the information needed to perform the feasibility assessment.

Feasibility is more than a judgment call. Unfortunately, inventors develop emotional attachments to their ideas. The idea becomes almost like a child in the eyes of a doting parent. I've seen lots of inventors with impractical ideas that totally lack feasibility, but they just won't let go and move on. Often, an impractical idea is only so because an enabling technology doesn't exist yet. Just look at DaVinci's helicopter or Tesla's turbine engine. Both of these we impractical at the time but today are common place because material and other technology was invented to enable their inventions.

Do the initial patent search yourself.

Fortunately, many ideas are feasible and worthy of moving forward.

Now that you have completed the project selection matrix, it is time to assess the feasibility of your selected project. Wouldn't you like to know that after the product is developed and sold, you could make a profit? (Of course that's after figuring out that the product can actually be designed and built.) In the spirit of "fail fast" we are going to perform an initial feasibility assessment here. This initial assessment is designed to filter out early in the development cycle the projects that are not feasible. There will be an opportunity later for an in-depth evaluation.

You might be asking right now, "Why don't I just hire out a patent search for my initial research?" That is a good question and demonstrates that you are thinking for yourself. Many inventors do this. However, what if you could do some work on your own to validate your idea and assess feasibility before spending the money for a patent search. The proper use for a formal patent search is in preparation of your patent submission and should really just confirm what you have already found. There is much more information in the public spaces that a patent search may not even identify. Finally, you really should not take anything someone else tells you at face value. In the

Ideation

The Practical Inventor

world we inventors live in, it's commonplace for people to throw wet rags on enthusiasm over a project. My general rule of thumb is that if, at first, my gut tells me something is not right with what I am being told, I take note and thoroughly check it out. Secondly, if it seems to really make sense, take note of it and still thoroughly check it out.

We have another exercise coming up in which we will formally assess your idea for an initial feasibility. There are four main areas to consider when figuring out if your idea is feasible.

The Practical Inventor   Ideation

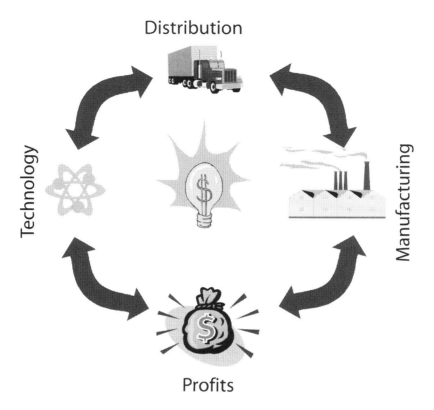

1. Distribution

How is your product going to be distributed to the end customer? If you are thinking of licensing, your licensee will want to know this. If you are thinking of catalog distribution for example, typical catalog markup is 3 to 5 times your selling price to them. Don't just think that someone else will figure this out later. You need to at least have some ideas to start with and it will not just happen.

2. Technology

Does the technology exist today to make your idea work? If you are unsure, take some time and see what is out there. I have performed many "Technology Benchmarks" for corporations for this very purpose. Don't overlook university research either. If you have answers to some of the hard questions, just write them down in your journal, you will most likely come up with an answer later. I don't think that lack of technology slowed DaVinci, Edison, or Tesla more than time to get a cup of coffee and think about it (at least I do the coffee thing).

3. Manufacturing

If you are going to sell something, you will need to build it. There are, as they say, "many ways to skin this cat". The question usually comes down more to which method to use and the answer changes depending on the annual quantity produced. The easiest way to determine if it can be built along with a relative cost is to look at similar types of products on the market.

4. Profits

Can you make any money with this idea? The normal answer is "Of course I can!" The reality part though, is often a lot more sobering. Most ideas I've seen can be profitable but it will take some work. The easiest ways to assess this without having a design is to just see again, what comparable products are available. If you have an idea for a new bicycle design and we know that bicycles are sold every day, you can logically assume you can realize a profit given appropriate Technology, Manufacturing, Distribution and of course, sufficient buyers. You should always question why they would buy yours instead of what is already out there.

Exercise 2-2: Initial Feasibility Assessment:

This assessment is more of a judgment call than an absolute verification that this is feasible and practical. This exercise is designed to help you:
- Fail Fast
- Identify more difficult questions you may need to answer later.
- Start developing a list of features and benefits for the product.
- Start thinking about different methods and strategies to accomplish your goal.

Create an Initial Project Plan

Now that you have completed the initial feasibility, what did you find? Your idea looks feasible and you are ready to continue or you are questioning moving forward. If you are definitely moving forward, congratulations and good work. If you found that there is something exactly like yours and no opportunity for you to improve on what is there now or your idea is just not feasible right now, it is better to find this out now. What is your next idea? If you found your idea is on store shelves or in use by NASA then you should be encouraged, you are on the right track. Remember, the thinking here is to turn your idea into something that can generate revenue for you, if someone is already doing it, they simply beat you to the punch, move on with another idea.

I presented the idea of a plan in the introduction. If you are at all like me, you would skip the introduction because you just want to get to the meat. (At this point, perhaps you should stop and peruse the introduction again). In this

> A plan gets you there faster and with less expense.

section, we are going to create an initial project plan and then fine-tune it further in chapter 5, planning. This approach will give your mind some time to process what you've put together and you will end up with a much stronger plan. Also, it is important to review and adjust the plan on a regular basis.

A project plan is a series of steps designed to move your idea forward in a methodical fashion to a desired outcome; in this case, a revenue producing product. The plan will sequence the development from idea to product in synchronization with your business development. This synchronization

Ideation   The
                                            Practical Inventor

ensures that you have everything in place, a good market for the product, and sufficient capital to properly apply for your patent when the rime is right.

You probably have some kind of a timetable that you want to follow. The plan will likely have more milestones without dates for now and that is OK. In chapter 5 on planning we will discuss some basic charting for your project. For more complete charting, you can hire a professional who specializes in helping people develop detailed and concise plans. I would recommend retaining the services of one of these people when the time is right – not quite yet.

The Practical Inventor

Ideation

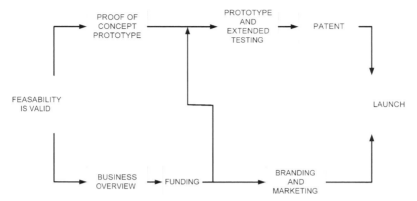

The chart on this page is an example of an obviously oversimplified plan starting at the point where you have determined your idea is feasible and are ready to move forward developing your product. Even so, a lot of project plans start with plans not much more complicated.

- First, mark down your starting point, where you are now.
- Next, mark down where you are trying to get to. In our case right now, let's say product launch. Leave plenty of room between these first two boxes.
- Work backwards from launch and identify the major steps that need to occur. Build paths backwards until you reach the point where you are now. It is likely that you will have multiple paths.
- Launch is the point where you are ordering your production tooling and getting ready to offer your product for sale. If you want to license your product, now is a point where you have good value. Branding and Marketing is critical. People generally will not buy your product if they do not know about it.
- Funding at this point usually takes the form of seed capital in unsecured personal loans from friends and family.
- A Business overview is a 10-16 page document that will use information you are collecting right now and will grow into your business plan when you are ready for it.
- Patents are not something to rush into at the start. Do your homework and be ready to do it right.
- Prototype and extended testing can be expensive. This prototype should be built to the production design. Proofs of concept prototypes are something you, as the inventor, should do yourself if possible.

Ideation *The Practical Inventor*

Exercise 2-3: Initial Project Plan:

Draw your initial plan in your design journal. Don't worry about having it perfect or in great detail as this is only the first pass at it. The intent of this exercise is to get you thinking about a plan. You start putting something, anything, down on paper. For now just start where you are on the project and create a diagram, or a flow chart, to list the steps you think are needed to take your idea from where it is now to the point where you can make sales.

If you are not sure about something, say so, we can address it later. If you miss something at this point, don't worry. This is only your initial plan. In chapter 5 on planning, we will discuss more fully the synchronization with the business development sequence and put much more detail into your plan. This exercise is intended to get you thinking along the lines of planning. For myself, I work much better if I am able to think about things in the background for a while.

Create a Product Specification

Now is the time to take off your inventor's hat. Painful as it may seem, you need to put on an engineer's hat for just a bit. If you are not an engineer, don't worry, it is one less handicap to overcome.

The product specification establishes important parameters for your product such as: size, weight, operating voltage, and specific functions or features. This is a data sheet on your idea and the product you are creating. Soon you will be hiring someone or a group to help you
move your product forward. The Design Specification will convey the specific details of your product to ensure the development moves in the direction you envision your product. It is amazing the number of people that have an idea and have companies formed; patents in process or issued, raised funds, spent countless dollars on engineering and even production tooling without a formal Product Specification. Such an approach doesn't allow you to move forward while expending *the least amount of time, money and effort* Let's not get to the point of paying for a production tool and making production parts only to find out that your product is missing critical elements that are still locked away in your mind!

Ideation

The Product Specification at this point is a proprietary document so treat it appropriately.

If your idea is a service or something less tangible than an actual "widget" that you are going to build, the Product Specification is still vital. It is the definition, the manifestation of your idea.

Phantom 1 Product Specification (example)

Electrical:

Operating System voltage: 12 VDC
Maximum charge current: 25 Amps
Trickle charge current: 500 mA
Full Charge voltage: 15.4 VDC
Charge cut in: 13.5 VDC

Mechanical:

Target size: 4" X 4" X 1.5"
Enclosure: flame retardant and UV stabilized ABS
Electrical Connection: Screw terminal barrier strip
Construction: encapsulated printed circuit board
Indicators: LED indicators for: Charging, Finishing, Source Ready, Power Divert

Target System Requirements:

Power source: up to 200 Watt, 12 VDC photovoltaic arrays or wind generator
Power storage: bank of 12 VDC lead acid batteries.

Operation:

1. When power source is ready, charge battery bank to the full charge state.
2. Upon full charge, switch to trickle/finishing charge and make the raw power source available on a power divert output.
3. Charge current is limited to 25 Amps max with short circuit fold back.
4. Charge kicks back on at the charge cut-in voltage.

Exercise 2-4: Initial Product Specification:

Write your initial product specification. I would recommend creating this document on your computer, print it out and paste it into your design journal.

Wrap up on Ideation:

Wow, congratulations! You have just gone far beyond where most people with an idea are willing to go. In this chapter you have accomplished:

- Picked one project or idea from a list of ideas.
- Performed your initial research and feasibility to determine this idea is still making sense.
- Figured out what you think are the next steps and their sequence as documented in your initial project plan.
- Completed the agonizing task of creating the initial Product Specification for your idea.

In short, you have picked a project that is practical and likely to make you some money, defined the product in detail, and set a preliminary plan in place to move forward with it. You should be very pleased with yourself. Take a break and reward yourself with something you enjoy. Personally I think a weekend at the Hot Springs in Glenwood Springs would work for me. This might sound frivolous but the reality is, your mind is going to be munching over everything you've just finished.

The Practical Inventor

Ideation

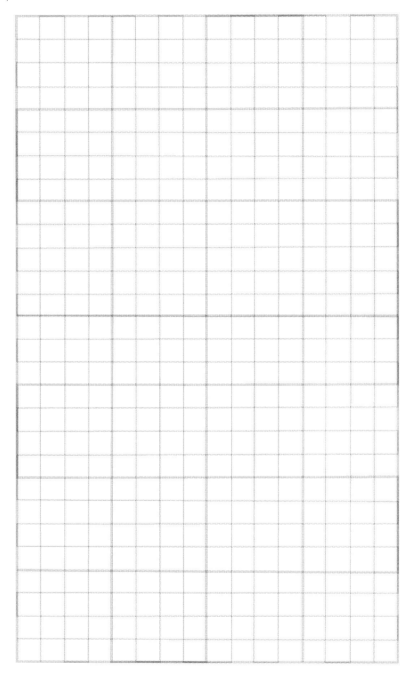

Chapter 3

Feasibility

Feasibility

At some point in the life of your idea, you (and only you can really do it) need to look critically at your idea and be brutally honest with yourself about its feasibility. If, after analyzing it, you feel positive about moving forward, take time to reward yourself on completing one of the less fun parts of being an inventor.

What if your honesty check reveals a project that may not live up to your expectations? Take a brief time to grieve the loss. Then shelve it. Perhaps you can resurrect it at a more favorable time or in a modified form. Yes, it's hard to walk away from an idea you've given birth to and believe in your heart to be good. Sometimes, however, there are excellent reasons for you to do just that - walk away and go to the next idea. Napoleon Hill in his book Think and Grow Rich showed how some of the people we consider today to be very successful people experienced failure after failure. They were on the brink of ruin before having a breakthrough. Remember this, successful people are those who pick themselves up once more than the people who were not successful.

The Practical Inventor — Feasibility

You may be asking yourself, "What is this word *feasibility*, and why are we considering it if I haven't even built my first prototype?" That's a good question. In response, I ask you, "Why spend time and money on something that isn't going to make sense?"

Feasibility is like a filter. You use it to stop activity that is not going to be productive. If practical, do this as early in your project as possible. Many years ago a good friend of mine coined a term and he used it to launch a highly successful program at his company. He called it "Fail Fast". This Fail Fast program was based on evaluating the feasibility of an idea and killing a project early to save time and money the project didn't meet certain criteria for probable success. This might sound like a negative program but it was actually positive. When you get past the Fail Fast stage, you have a higher confidence level in the feasibility of your idea.

Feasibility should be evaluated on its merits in three areas:

1. Technical feasibility
2. Market feasibility
3. Practical application.

Feasibility   *The Practical Inventor*

Technical Feasibility:

The main issues with technical feasibility are:
- Does the technology exist for your idea to function properly once it is built?
- Does the technology exist to manufacture a product based on your idea?

You may be thinking "Duh!!! That's obvious!" and you should be right. But it's amazing how many ideas, including patented ones, could never be produced due to the technology limitations. Sometimes there are material limitations as in Nicola Tesla's turbine engine or as in early NC machines that used 2 circuit cards to create what is available in a single 8 pin IC package today. If your idea is not technically feasible, it can still be a worthwhile project. Something must be developed or a material created to enable your product. Some examples are:
- Material does not exist – A material must be engineered, created, or a substitute method developed. What Tesla needed for his turbine was titanium rather than steel.
- Material handling methods do not exist – A method for safely handling the material must be developed or alternate material found. In the Manhattan Project, they were developing the material for the core of the atomic bomb. No one had ever handled highly processed Uranium before and there were no established techniques for doing so. They had to develop methods as they went. Since that time, a whole industry has emerged for handling dangerous materials.
- Manufacturing method does not exist – A new way to cast, mold, machine, assemble, etc must be developed. Tesla also had a problem with his turbine also in the manufacture of his blades. The technology has improved for casting metals with a high degree of precision. However, opportunities still exist in this process to reduce the cost. New laser technology that could render the casting process obsolete and has already enabled manufacture of complex parts that previously would have been very expensive or impossible to build.
- The infrastructure does not support your product – The infrastructure must be created or an alternative design for the product must be created. For example, the electric vehicle can't feasibly be used for distance driving because there is no good method for recharging the batteries. The solution a few creative people found was to put a small gas or diesel engine and generator that would start and recharge the batteries when needed. There has been talk about stations to swap battery packs much like you swap propane tanks today.

The Practical Inventor

Feasibility

When you have an idea that turns out to be technically infeasible, that just means that you are probably a little ahead of your time. Don't worry; it didn't stop Leonardo DaVinci from sketching up his helicopter. It may mean however, that you pick a different idea to concentrate your efforts while you watch

> Failing fast saves money and allows you to move to a better project sooner.

for an enabling technology or ideas that could make the idea technically feasible.

Exercise 3-1: Technical Feasibility

In the next chapter you will perform a technology benchmark. During this exercise, you should uncover information that you will find useful. Then, be sure to bookmark websites and to write down not just information but sources of information.

1. Make a list of different technology areas that you will need for your product.
2. Make a list of every product and technology that you find that is similar to or related to something you will need for your product.
3. Identify the technology areas that may be a challenge and explain why.
4. List any technology areas that, if something existed, could facilitate better performance of the final product or reduce the manufactured cost.

Feasibility

The Practical Inventor

Market Feasibility:

Market feasibility includes issues and concerns that could facilitate or inhibit your ability to sell or distribute your product. Some possible issues are:

- Benefits are not obvious. When people look at your product, the benefit of your product over other products may not be apparent. When horseless carriages first appeared, most people looked at them with disdain because there were no roads, gas stations, and they were not accustomed to dealing with what was then an unreliable machine. They couldn't figure out what was wrong with their horse?

- Too expensive. Even if nothing else like your product exists and consumers can readily see and appreciate the benefit, there is a limit to what they will pay for it. When color TV sets first hit the market, they were marvelous. People would stand outside a store window and watch whole programs just to see them in color. However, the TV sets were very expensive and most people were not willing to pay that price for color.

- Other competing products. You may find your product already on the market or learn it was on the market 20 years ago. You may find another product on the market that fills the need, but yours is different. For example, you may have developed a new fishing reel that is enclosed and you can cast with only pressing a button on the reel with your thumb. What is on the market now is a fishing reel where it is hard to cast, the string is all exposed, and can easily come off and get tangled.

It comes to the question of why someone would spend money for your product over something else. Don't forget, that something else could be shoes for the kids.

Exercise 3-2: Market Feasibility

1. Review exercise 0-1 where you defined what you idea is in terms of benefit to society.
2. Refer back to your findings in exercise 3-1 Technical Feasibility.
3. What other products did you find on the market that address your same societal issues and solve the problem, even partially?
4. Regardless of competing product, is your product something people would be willing to pay your retail price for?
5. Create a survey and conduct a survey. You can make questions hypothetical without disclosing what you are doing. Ask questions that would provide actual customer answers to the questions listed previously in this exercise. Record the number of people you talk to and their answers.

Practical Application

This is the part that is not so easy. Based on the two previous exercises, you either do or do not have technical and market feasibility. This portion of your feasibility analysis, however, looks inward and solicits objective feedback from a close circle (hopefully you have a team helping you). What it comes down to is answering the question "Is my idea practical?" That is almost like asking a mom if her baby is beautiful. I remember in the pre-Windows computer days (this could be dangerously revealing my age – I was very, very young then) some people wanted to do everything on the computer. My rule was: if I could do it faster and as good on paper, use paper. Thanks to those who didn't follow my rule, I do almost all that stuff on the computer now too. There can become a time when the impractical becomes practical, and sometimes the impractical is always impractical, like freeze-dried Ice Cream?

Exercise 5-3: Practical Application

1. Make a list of all the different ways people deal with the issue today that your product will address. Don't forget "ignore" and "just live with it".
2. Cross out the items on the list that cost more money than your product and that yours will replace because it is as good or better.
3. Underline the items on the list that your product will replace but your product costs more money.
4. Circle the items on the list that address the issues better than your product will and you will not replace.
5. Look at your list and honestly look at the market viability of your product with this information. Does it make sense?

The
Practical Inventor   Feasibility

Decision Point

You have taken a hard look at the feasibility of your idea and product. This is a tough process and you are to be commended for making an honest effort. There are two possibilities at this point. First is that you have determined there are no issues with feasibility and you are ready to charge ahead. Do so! However, if you found feasibility issues with your product, you have three choices:

1. Drop the idea and move on to your next idea.

2. Table the project and move on to your next idea while working out ways to resolve issues that have been identified.

3. As John Paul Jones said, "Damn the Torpedoes, full speed ahead!" He was also an entrepreneur, you can tell by his attitude. If you have good reason to believe your idea is sound and you've uncovered some obstacles you can deal with along the way, get to it.

Exercise 5-4: Decision point
1. Did you find that you have issues or are you free and clear to move forward?
2. If you found issues what are you going to do?
3. Write out your action plan for each item to make each of these a non-issue. Include who has what responsibility. When is your goal for having it finished?

Chapter 4

Intellectual Property

Important Notice:

None of what is presented here should be construed as legal advice, nor is it intended to replace a good IP attorney. There is no substitute for appropriate professional advice. The basic information on IP in this chapter is intended to help you prepare an IP strategy sp you can start moving forward. This chapter will also help you to prepare to meet with an IP attorney to save you both time and money when the time is right to take this step.

Intellectual Property

Intellectual property (IP) is an area greatly misunderstood by inventors. Most inventors think immediately about getting a patent to safeguard their rights. They seem totally unaware that other methods exist to protect their ideas that, depending on the circumstances, may be a better choice. In this chapter you are going to learn about the sequence and various methods of protecting your idea. Also, what is meant by protection? This chapter is not an exhaustive work on IP, but you will learn the basics and more importantly, how to evaluate and properly use the tools you have available.

The Practical Inventor
Intellectual Property

What is IP?

Intellectual Property (IP) has two aspects. First is the proprietary information that defines how the public understands and sees who you are. This IP is your name, logo, jingle, smell, sound, etc. Your identity is protected with trademarks, service marks, trade names, and branding. We are not going to cover anything about this type of IP but you can read more about it in a book by David Pressman called "Patent it Yourself" (see the appendix for more information).

The other aspect of IP is the special knowledge you posses that helps your business. This knowledge can take the form of a product idea, design improvement, process, formula, recipe, text (such as a book), photographs and artwork, computer program, etc. These types of items are protected using patents, copyrights, trade secrets, and sometimes your cousin Gino and his boys.

The copyright office website is http//copyright.gov and the Patent and Trademark Office website http://www.uspto.gov. Both contain is a wealth of free information. I would highly recommend that you take time out right now and spend some time poking around these websites.

Copyrights

The copyright is intended to protect specific creative work. This work includes things like computer programs, written text, photos, and graphics. The tricky part can be the ownership. Let's say you are writing this book. You can copyright the book and everything in it establishing irrefutable ownership of it. However, if you hire a graphic artist to create a snazzy cover, who owns that artwork? Normally, the artist who creates the work owns it, even when you hired and paid them.

Intellectual Property   The Practical Inventor

Any time you hire someone to perform some type of work like this for you, whether working as an employee or and independent contractor be sure to have a written contract that includes a clause that stipulates it as a "work for hire" or assigns all creative work to you as the owner. This can be a very simple statement assigning all ownership to you but it is critically important. In fact, it is a good practice to develop some template contracts and have an attorney review them to ensure you have your bases covered.

Back to my book example: the copyright for my book is good for 70 years after I as the author die. With work for hire, the copyright is good for 95 years from the first publication or 120 years from its creation, whichever is first.

One interesting aspect of a copyright is that you do not have to file for it. When the work is created, you can declare a copyright by using the © symbol date and who is copyrighting the work (example: ©2007 Veritek, LLC). Registering your copyright is easy and inexpensive on-line at the copyright office. The main benefit to registration is
Comes in the ability to prosecute when there is a copyright infringement.

> Always have a "work for hire" clause to retain your ownership.

What is considered fair use and what happens when someone infringes on your copyright? Others can use your copyrighted material for educational purposes and sometimes for non-profit type work. Also, parodies can be allowable but can come into infringement of trademarks you have established. Fair use of a copyright can be narrower than you might expect. To determine if an infringement has occurred, the courts look at several aspects: how someone else's use has impacted your market, how much of your work was used, and how this type of material would normally be used. They also consider the degree of infringement. However, getting to this point means that you as the owner of the copyright, must take the suspected infringing party to court - yes sue them. If the defendant is found guilty of infringement, the court can issue an injunction to stop further use, order the seizure and destruction of the infringing material, and award you attorney fees, compensation, and damages.

The Practical Inventor   Intellectual Property

Patents

A patent is a public disclosure of your technology, defining the specific area of the field where you have a monopoly for a period of time. The intention is to provide a window of time where you can have this monopoly to recoup your investment in developing this technology and to make a profit. It is also a facilitator to "selling" your idea by licensing it to a third party - someone else or another company. You can however, license your idea without a patent.

Anyone can file for a patent; you do not need an attorney. But why would you want to? I suggest you find a reputable patent attorney and leverage their experience and expertise. If your idea will become a product with good market value, hiring a good attorney is not an expense it is an investment. The real question is not whether but *when* to hire an attorney in the sequence of your development cycle? My answer to you is "not yet". We will look at this later when we talk about strategy. For now, no public disclosures (this includes one on one meetings at restaurants) and have signed non-disclosure agreements from everyone.

> Don't rush out to file your patent. Patents happen in proper sequence.

Intellectual Property

The Practical Inventor

There are 3 types of patents:

1. **Utility** patents, the most common type, are good for 20 years. A utility patent covers functional inventions like a new door handle, electronic circuits, a drug, or anything that has a function. Also, remember that utility patents have maintenance fees at 3, 7, and 11 years after issue to keep it enforceable.
2. **Design** Patents are used to protect products with a specific design and are good for 17 years. This patent would cover things like a very specific looking floor lamp or a specific building design. The old McDonalds buildings with the golden arches running through the building or a Frank Lloyd Wright house would be examples.
3. **Plant** patents are good for 20 years and are used to protect plants such as grafted flowers or hybrid types. The plants can also be protected with a utility patent.

Patent filing deadlines are also something to keep in mind. What most inventors don't realize is that it is not critical to rush out and get your patent filed. In fact, rushing into it can be detrimental unless you just want a patent and are not worried about making a profit for your expenses. The reality is, you never have to file anything with the Patent Office while you keeping the disclosures in the private venue. When you are getting ready to launch or make any kind of a public disclosure, you want to have a filing with the patent office. This filing can be a Provisional Patent Application (PPA), or a full Patent application. While the US patent office allows up to a year grace period after public disclosure, there is no grace period with other countries such as with European patents. If you make a public disclosure without a filing at the US Patent Office, you can never get a European patent on your product.

A note about provisional patents from the Patent and Trademark office website: "Provisional Application for a Patent

Since June 8, 1995, the USPTO has offered inventors the option of filing a provisional application for patent which was designed to provide a lower cost first patent filing in the United States and to give U.S. applicants parity with foreign applicants. Claims and oath or declaration are NOT required for a provisional application. Provisional application provides the means to establish an early effective filing date in a patent application and permits the term "Patent Pending" to be applied in connection with the invention. Provisional applications may not be filed for design inventions.

The Practical Inventor
Intellectual Property

The filing date of a provisional application is the date on which a written description of the invention, and drawings if necessary, are received in the USPTO. To be complete, a provisional application must also include the filing fee, and a cover sheet specifying that the application is a provisional application for patent. The applicant would then have up to 12 months to file a non-provisional application for patent as described above. The claimed subject matter in the later filed non-provisional application is entitled to the benefit of the filing date of the provisional application if it has support in the provisional application. If a provisional application is not filed in English, and a non-provisional application is filed claiming benefit to the provisional application, a translation of the provisional application will be required. See title 37, Code of Federal Regulations, Section 1.78(a)(5).

Provisional applications are NOT examined on their merits. A provisional application will become abandoned by the operation of law 12 months from its filing date. The 12-month pendency for a provisional application is not counted toward the 20-year term of a patent granted on a subsequently filed non-provisional application which claims benefit of the filing date of the provisional application."

Here are the rules: Have a signed Non Disclosure Agreement (NDA) with anyone you show or talk to about your product or idea. Document the meeting in your design journal including where the meeting took place. Finally have no meetings in public places. Get your coffee, visit, talk business, but go to the car to talk about the product or idea.

Intellectual Property   The Practical Inventor

Trade Secret

A patent is a public disclosure (a patent in a public document) of your technology and establishes the scope of your technology (usually in the claims). The master art of patents is in disclosing enough information to obtain a meaningful patent while not disclosing enough key information to slow people down from copying your work exactly. Someone knowledgeable in the field should be able to understand what you are doing but likely not recreate your technology as well as you. With a trade secret, you simply don't tell anyone how the core technology works. This could be that only one person has the formula or that the technology does not leave your facility. A large materials and adhesive company does not even let their own sales staff into the manufacturing facilities due to trade secret information.

The key to trade secrets is:

1. Properly documented in your journal.

2. NDA's with anyone who will see it (document the event in your design journal)

3. Trade Secret information is clearly identified to all employees.

Trade Secrets never expire.

4. Trade Secret disclosure covered in employment contracts.

5. All employees working with Trade Secrets are regularly trained on the safeguarding of the Trade Secrets and documented in their personnel folders.

6. Breaking the Trade Secret information into different levels of security and disclosure where the absolute core of the technology is top-top secret. Think of the Military and their security classifications system of Confidential, Secret, and Top Secret. Also, while you may have the clearance level, there is still the need to know aspect.

Don't make things too complicated but think about how critical the Trade Secret information is. Would a leak cause an inconvenience or hardship? If the competition got a hold of it would it devastate your company? Your answers should tailor your appropriate level of protection.

Remember that most of us want to make a profit from our ideas, so if you keep your Trade Secret too secret, can you make money?

Some Examples of trade secrets are:

- Business Plans
- Recipes and formulas
- Customer lists
- Manufacturing Techniques

Work for hire agreements

Intellectual Property   The Practical Inventor

Exercise 4-1: Determining your IP path.

Use the following chart to determine your IP path. Most inventors using this manual will follow the Utility Patent Route.

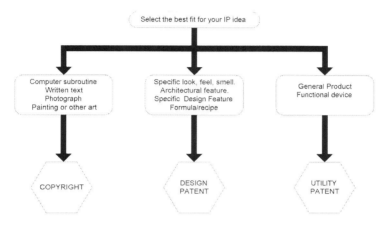

Document your findings and decisions in your design journal.

The hard part is deciding if you are going to keep your computer subroutines, formulas, recipes, or key information as a trade secret or patent/copyright it. This might be a blend of trade secret and patent/copyright.

Make an inventory of your IP for this project. This will be a two-column list in your design journal. On the left, list each specific IP and on the right column, the path for each item.

The Practical Inventor
Intellectual Property

The following list is an example:

client - server authorization	trade secret
cabinet design	design patent
feature character	trademark and design paten
method of game play	utility patent
method of accepting payment	utility patent
system architecture	utility patent
game card generator	trade secret
game chair	utility patent
door opening mechanism	utility patent

Good work. Don't worry too much about what to do next we will get there. For now, keep documenting work in your design journal.

Valuing IP

Intellectual Property is a tangible asset of your company but the value can be hard to assess. Ultimately, the value of your IP is the willingness of someone to pay money to purchase your product.

There are several reasons for valuing your IP. The foremost reason involves funding. You are talking to potential investors, and your best chance of raising the value of your proposal to the investors is to have a higher value on your IP. The question is how to value it. For example, I collect certain antiques. When selling them, even though there are clear blue book values published with specific detailed methods for rating condition, it still comes down to what the buyer is willing to pay. That same principle applies to the value of your IP; it comes down to what the investor is willing to pay for their return (not potential return).

What can affect the value of your IP?

- Prior rounds of funding establish a baseline of what investors paid in the past. Presumably, you have moved forward with company and product development so not only is their risk lower but you can charge more for their funding in higher value shares or lower interest rates.

- Your track record will drive value to your IP if it is successful. Consider this, if you had a car company and it was in trouble would you seek to hire Iacocca or DeLorean? Who has the stronger track record? OK, you might still hire DeLorean or someone with his experience but you will only pay him $100,000 a year rather than Iacocca's 4 or 5 Million dollars a year. *Your track record is your personal IP and people invest in it more than your prototype.*

- The strength of your team and their track record will drive value. This is your real team, not your paper team. I see a lot of startup companies list out their team with well-known, incredible professionals. The companies with higher value are the companies that have these team members engaged in moving the business forward, not just listed on their portfolio. (I am using business a lot in here because if you are looking at valuing your IP, you are talking business.)

- Sales - yes sales - increase IP value. Even in early stages, it is possible to take Letters of Intent and Purchase Orders for delayed product delivery. Another form of sale would be a contract for delivery of service based on your IP. These are all excellent for establishing ironclad value. You are establishing what people are willing to pay for product based on your IP and this is a commitment, not a market survey.

- Prototypes can increase the value of your IP. However, the closer you are to having a production design and production representative pre-production unit, the easier it is to establish a value as well as the higher value and lower risk consideration.

- The proposed business relationship causing the IP valuation can also affect that valuation. The reason for the valuation should be explicitly spelled out as to licensing, exclusivity, time frame, geographic areas, or revenue participation to name a few. An couple contrasting examples:

    a. The IP is likely to be valued lower if you are discussing a licensing to a specific market. However, the value can go up if you are discussing exclusivity.

    b. The IP is likely to be higher if someone is negotiating to buy your company or the IP completely.

You can optimize the value of your IP with some simple but effective techniques:

- Minimize initial cost outlays

- Reduce the cost of failures

- Increase the likelihood of success (stack the cards where you can)

- Reduce the likelihood of failure (remember the risk analysis?)

- Increase your rewards for success

The bottom line is that there is no one clear-cut method of establishing value on your IP because it is an intangible item. Look at the difference in value between Budweiser's trademark and Left Hand Brewing Company's trademark. How would you even start establishing a value? The following is a listing of some of the tools that could be used in establishing a value for your IP:

- Marginal Contribution looks at the ratio of your profit to overhead. Generally this is calculated as Earnings Before Interest, Taxes, Depreciation, and Amortization.

Intellectual Property
The Practical Inventor

- Cash Flow methods look at the opportunity cost for the investor against other potential investments and the risk associated with investing in you. (Remember the discussion on track record earlier) This method is more subjective because the investor will rate and discount your IP's ability to generate cash flow in a given time frame against other opportunities.

- Lost Income method looks at a hypothetical scenario of what would be the investor's lost income if they did not have your IP in their portfolio.

- Incremented Marginal Revenues looks at the contribution the IP can make and the cost for it to do so based on branding strength. Obviously, Budweiser will need less money than Left hand Brewing because of the strength of their brand.

- Residual Profits method attempts to isolate the profits specifically attributed to this IP. Often there are several IP's interacting to generate the profit. This method only works if you have identified and modeled the other IP's.

- Comparables method is like real estate valuation where you take a look at what other similar properties have sold in the area. If you can find a similar type IP that has sold in a business model similar to yours, this is a good baseline value for your IP.

- Cost Avoidance method looks not only at the revenue the IP could generate but also at savings that can be realized internally as a result of your IP. If you think about a 20% profit margin, every $20 saved as a result of your IP equals $100 in sales plus your income tax savings.

- Monte Carlo Simulation is used in many business and scientific scenarios. This is a simulation of your business as a result of the IP based on random occurrences. When you have something with established probabilities, it is relatively easy to set up an Excel spreadsheet with the probabilities and run thousands of events in the simulation.

OK, confused yet? No problem, at the point where you need to understand this, you should have a Chief Financial Officer on your team to make sense of all this for you.

The Practical Inventor

Intellectual Property

Exercise 4-2: How much is it worth?

As you have seen from the discussion in this chapter, determining the value of your Intellectual Property is not quite like figuring out the value of your old Chevy. In fact, you cannot yet accurately complete this exercise because will need information from other sections in the manual and team input for the best results. The goal for this exercise is to begin getting an idea for the value of your IP. The first step is to complete the IP scorecard. This was adapted from Karl Dakin's seed capital scorecard. Karl is an expert in technology transfer. His contact information is found in the appendix at the end of the manual.

1. In the table below, score each row in the Points column as appropriate.

Question	Value	Points
Out of 100 people, how many will quickly understand the value of your product and act to buy it?	1 point if less than 10 2 points if 10-20 3 points if 30-40 4 points if greater than 40	
Of your targeted customers, how many can you identify by name?	1 point if up to 1% 2 points if 1 to 5% 3 points if 5 to 10% 4 points if more than 10%	
In small quantities, what can be the profit margin on the sale of a single product?	1 point up to 10% 2 points 11-15% 3 points 16–20% 4 points over 20%	
How many other products are on the market that offers the same benefit?	1 point if less than 20 2 points if less than 10 3 points if less than 5 10 points if none.	
Score in this row is based on of you or a team member is a functional expert in the field of your product. This is a measure of the strength of your team specifically in the market and technology area that you will be selling your product into.	1 point if no expert or experience in the field. 2 points if limited experience (up to 5 years) in the field. 3 points if extensive (5+ years) experience. 5 points if a recognized expert in the field.	
Score in this row, your overall team strength. This row should be completed after your product plan is in place (product planning chapter). You are not a one-person show.	1 point if none of your team is in place. 2 points if 25% of you team is in place. 3 points if 50% of your team is in place. 4 points if 100% of your team is in place and functioning.	
How easy would it be for someone to copy what you are doing?	1 point if this is so easy to copy that you only have to see it to copy	

Intellectual Property

	it. 2 points if your product can be easily reverse engineered if they have one to copy. 3 points if there would be a considerable amount of investment and engineering to copy your product. 10 points if you have proprietary or Trade Secret information that without it, there is no way they can copy your product.	
How for along are you on the product development.	No point if you just have the idea and it is not properly documented (see the chapter on documentation) 1 point if you just have the idea and have it documented properly. 2 points if you have a working proof of concept prototype. 3 points if you have a good working model but more developed than a proof of concept. Something close to production representative. 4 points if you have a production representative model. 10 points if you have engineering and assembly drawings ready to release for manufacture.	
	Total	

2. Add up the Points column and enter it into the Total box.
3. For the multiplier below enter a 2 if you have current sales of a similar product or if you have already received partial investment funding. Enter a 4 if your program is fully funded internally either by profits from sales of other products and services or through your own means without investment funding or loans. Enter a 1 if neither the above is true.

  Multiplier = _____

4. Multiply the Points total by the Multiplier. This is your valuation score.
  Score = Multiplier X Points
  _____ = _____ X _____
5. Take a moment to reflect on your score. The maximum you can get is 204 points. The higher the point score, the more credibility you will have with whatever value you assign to your IP. 7 points is the lowest score. If you score less than 20, your need to ask yourself if this is what you should be working on.

6. As you move forward, repeat this exercise and work on areas that are lower to bring the points up.
7. A score of 25 to 35 is very good and commendable before the multiplier and likely a strong product.
8. In the chapter on Feasibility, you evaluated market viability. At that point you gained some idea of the market potential (in annual sales) for you product. Put that figure in here.

   What is the annual sale forecast for your product?
   $_____

   Based on your figure from above, multiply your sales forecast by the Sales Multiplier and put it into the annual sales value box:

Score from above	Sales Multiplier
0 to 15	0
15 to 20	0.1
21 to 23	0.2
24 to 30	0.3
31 to 50	0.5
51 to 100	0.75
100 to 200	1
Above 200	2
Your tentative annual sales value	

9. Companies with established sales are valued on a factor that takes into account the amount of sales, stability of the company and market, and the maturity of the market. The company valuation is often based on this factor times the annual sales. For a stable company in a mature market this factor is 2 but in a growth market, it can be 5 or higher. You job is get this factor as high as possible and to convince whomever is looking at your IP value that all of the numbers you've come up with in this exercise are valid and to get this factor as high as possible. For now, we are going to use 5 and let them argue otherwise.

   IP Value = tentative annual sales value X factor

   IP Value = _____ X 5 =
   $_____

Remember that the valuation is still very subjective and that everyone you talk to is going to have a different idea. Do not get your feelings hurt when someone values your IP lower than you. Use the tools you've learned here to educate others about why your IP is valued as it is, discuss it, and come to a reasonable agreement as to the value. I confess that my IP valuation is always much higher for my ideas than what others see it, at least during the beginning of a discussion.

Protecting IP

Up to this point, you have documented your idea, determined what method of "protection" to use, and determined a tentative value for your IP. What are you going to do with this information? Ideally, you will continue on with this chapter and keep completing the exercises. The goal is to move your idea and product forward continuing on with the exercises should accomplish that for you.

Once again, let me remind you that nothing said here should be considered legal advice. Rather we are talking strategy, timing, and important points that will help you move faster and save considerable money when you do start working with an IP attorney.

You are now going to decide on a strategy for protecting your IP, create a policy to document your strategy, and do it. Step one is determining your strategy and since I broke my crystal ball last week, I know nothing about your idea or product. Therefore, this discussion will be more generalized than I would like. Regardless of what you have decided to this point, your decisions should be brought back to the money (and safety).

Whatever IP protection method you select, you need to realize that the concept of protection is a bit misleading. When someone infringes on your IP, the Government does not swoop in and defend you. The burden is on you (or possibly your licensee if you've licensed it) to sue the infringing party in court. Remember to factor in the lost business due to the time spent defending your IP in court and that the court and your attorney are burning your hard earned cash until and if you win. Don't be scared off by this thought. Rather, consider the business and economics of your idea before taking a side trip from your intended route toward success and profit. Spend your limited resources wisely, when and where appropriate. Remember that if your idea is really good, others will copy it and you will have to defend your IP.

Copyrights

Your very first step would be to add the copyright symbol, date and name to the bottom of the document or to proclaim it as copyrighted material in the header section of the text. If you feel there is sufficient commercial value, to register the copyright with the Copyright office as discussed earlier, the filing fee is currently $45.00.

Patents and Trade Secrets

The first mentality you should adopt is that all of your ideas are Trade Secrets. If at some point in the future you decide to file a patent, then that disclosure will nullify your Trade Secret on the portion of that is disclosed in the patent. My general rule is for information that is critical to my product or company's survival will always remain a Trade Secret or at least the critical element of the idea. You can do this when you have something more complicated and surround it with different utility patents. For simpler products, a patent or group of patents will suffice.

The key to remember is that the getting patents cost money along with maintaining and defending a patent. Only you can decide if there is really sufficient commercial value to make that investment. If not, think about

Intellectual Property — The Practical Inventor

running hard, making a wad of cash and walking away at some point in the future.

The basics for protecting your IP are:
1. Proper documentation - we discussed that in chapter 1 on documentation

2. Use a Non-Disclosure Agreement with everyone before disclosing any information.

3. Keep your venues for disclosing information in private spaces. Telling someone about your idea at a table at a restaurant or coffee shop is not private.

4. Everything that is part of your IP or proprietary information needs to be identified clearly. You can by a stamp with red ink that reads "Confidential" or "Proprietary and Confidential" and stamp the top and bottom of each page of paper. Electronic files can have this in the header and footer.

5. Record in your design journal who, what, where, and when any of your IP was disclosed.

6. Be prudent and diligent, not paranoid.

7. A patent is not automatic protection; it's only as strong of protection as your ability to hire an attorney and take someone to court when they infringe. Your attorney may be able to resolve the infringement with just a letter. Some people (or companies) will infringe on your IP, hoping that your patent pending does not come through to a full patent or that you won't take them to task. A portion of them, when confronted, will negotiate a licensing and royalty agreement with you, but many won't. You'll have to take them to court. The Government or Patent Office will not do this for you. Also, you cannot take someone to court with your patent pending, which only scares off the honest people. Your full patent must issue before you can sue the infringing party. You can, however, ask your attorney send them an infringement notification letter with the patent pending and sometimes negotiate a license with them even though you do not have a full patent.

8. When talking with most people about your idea, discuss features and benefits rather than how or what. Most people only need to know what it will do for them. They ask, "How are you doing that?" An appropriate reply is that it is proprietary information and you are not at liberty to divulge it yet. The line, "Well I could tell you but then I'd have to kill you." is greatly overused and normally inappropriate (though fun to say).

> The first mentality you should adopt is that all of your ideas are Trade Secrets

9. Your employees and team members, even your advisory board or board of directors, must be trained on proper security of you IP and how you have it identified.

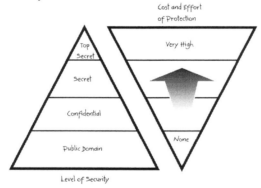

10. Remember, the higher level of security, the more difficult and expensive to properly control. Divide the IP into different levels of security, perhaps including::
    a. Non-classified material such as features and benefits, sales information, information that is public knowledge and related.
    b. Confidential material or information that would be inconvenient if it got out but not critical. This could be information that would tip off your potential competitors about what you are doing with more detail than you wish to share.

Intellectual Property

    c. Secret material is specific information on what you are doing or how you are doing it. If this information got out, it would seriously hurt your profitability or ability to dominate based on the technology. Improper disclosure of this information would have serious implications but you would still be able to move forward and to compete in the marketplace.

    d. Top Secret information is the key, critical elements that, if improperly disclosed, could or would completely squash your ability to move forward. Even though you probably have people on you inner circle that you would trust with your soul, there is no reason to share this information even with them unless they need it in the performance of their duties.

11. Become a good judge of character and develop a relationship with someone before disclosing information. Everything we have discussed works with honest and ethical people. People will take your ideas; it is all part of the game. Do not let this stop you from moving forward, just be aware that if you really have a good idea, people will copy and infringe. You can control this to a certain level by getting to know people you are disclosing to better before disclosing the more critical information.

> A trade secret can be more valuable than several patents.

12. Finally, in the words of Bernie Doorman from CEO Space (CEOspace.biz), "In God we trust, all others do contracts."

The Practical Inventor
Intellectual Property

Exercise 4-3: IP protection policy

Go back to the IP inventory list from the first exercise in this chapter. You will likely want to return to this exercise after going through chapter 5 on project planning. You may have identified items for patent, but you will probably not be starting into the patent process yet. In fact, you may decide to skip the patent, file a provisional patent, or file the formal patent. It has often been said, timing is everything. This applies to patents too. Even if you decide on the patent route, it may not be time yet for your filing. Think about economic value, cost and time.

Non Disclosure Agreements

The Non Disclosure Agreement (NDA) is your tool for establishing with people that you are providing to them certain proprietary information along with the rules for your identifying the IP, disclosure, their safeguarding it., and surrender of material should your relationship with them end.

Some rules for your NDA processing and care are:
1. There are 2 copies, one for you and one for the other party.
2. All pages should be initialed and dated by both parties.
3. Keep your copies in one folder. I scan them all and keep them as pdf files in a folder in addition to the paper.
4. Record the NDA in your design journal. (date, who, where, what was the nature of the discussion or disclosure)
5. Execute an NDA before disclosure.
6. After you have the NDA, the information must still be disclosed and discussed in private venues rather than in public places. This means that if you are meeting in a restaurant or coffee shop, these locations or anywhere that you conversation could be overhead or notes seen by the general public, it is not a private venue. One example of a safe disclosure would be to meet at the coffee shop, execute the NDA, discuss non-proprietary items, and then move to your car for the disclosure and discussion around that.

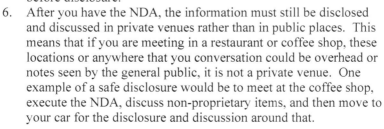

OK, just record everything in your design journal.

7. Finally, be very clear about the material that is subject to the NDA.

> Keep discussions about proprietary information in private spaces.

Identify each item of information, clearly mark it, and document it in your design journal.
8. This agreement is only as strong as the integrity of the person you are disclosing the information to. Having the NDA still does not mean you should lay your soul bare. Only disclose what is necessary for your immediate purpose and do not be afraid to tell them when you are asked a question that what they are asking is not necessary knowledge for the task at hand.

Exercise 4-5: Your NDA

Prepare a Non-Disclosure Agreement. Typically, the best bet is to hire a patent attorney to do this for you.

Planning your patent work:

Patent work is very expensive. Throughout this chapter and manual, we have been talking about different methods of protecting your invention and methods of moving forward with the development and commercialization of your invention. Because this is such an important point, it is worth taking another look at:

- The first action is documenting your idea properly in a design journal as described in Chapter 1 – Documentation.
- Keep all your disclosures within the private venues using NDAs and documenting in your journal

- Do your homework. Put your business in order and raise the funding necessary to complete a successful launch or licensing of your inventions. This stage is often referred to as seed funding.
- File a provisional patent before any public disclosure. You must have your business put together enough so that you will be able to pay for your patent work within 12 months. Talk patent attorney for a ball park cost. Typical patents cost $10,000 to $20,000, but I have seen patents cast as much as $250,000.
- File your patent applications within 12 months f filing the provisional patent. This includes your international patents.
- Once your patent is issued, all products covered in the patent must be marked with the patent number. Failure to do this can impair your ability to prosecute any infringements.
- Don't shop cost for a patent attorney. This is not an area to scrimp on cost. Choose a patent attorney who has experience and a good track record in the industry area of your patent with reasonable technical expertise. I have seen people lose on patent infringements over one very small omission in a claim.
- Remember to include your patent maintenance fees in your planning and budget.

Patents and Technology Searching

Patent searches are necessary as you move forward. Having a formal patent search completed by a professional is necessary when putting together your formal patent application. While we will be addressing patent searches, this section is not specifically about patent searches. In fact, you are much better off in many respects if you learn to search for yourself, especially at the start of your project.

A Technology Benchmark must be performed by the inventor for the greatest benefit. I've performed these for large corporations for years and unless you've got deep pockets, probably you're far better off doing it yourself. Besides, there is a larger purpose than money for you as the inventor to do the Technology Benchmark. Some of the benefits from performing your own Technology Benchmark are:

1. It can be a lot of fun.

2. The information gathered will form the basis of your feasibility analysis.

3. You will gain new ideas to potentially enhance your idea including features, applications, spin-off products, and additional markets.

4. You gain first-hand knowledge of the market you will be competing in.

The Technology Benchmark is like a detective game in many ways. The patent search part is so simple it is almost boring but the rest of the benchmarking is going to require that you think outside the boundaries of normal searching (something inventors are usually good at).

The Practical Inventor
Intellectual Property

The Technology Benchmark includes searching and evaluating the following areas:

Area	Pros	Cons/Limitations
Patents	Very specific and detailed information is available on published patents.	There is a blackout period after the patent issues where you can't find anything about it. Also, Provisional Patents are not available to the public so you have not idea of what is in process.
General Public Domain and Public Market	Generally easy to find information on the web.	Specific details are often sketchy.
University, Government, and Public Research	• Detailed information is often available. • Universities are eager to publish. • Most contract research has publication requirements as a condition of the contract. • Publication is one method they use to validate their existence.	• Need to know the Universities with research along your specialty. • Fee based search engines may be required to find the information.
Private Industry Research	If you can find any, it validates the potential market value for that area of industry.	Pretty hard to find and really need a strong network as this information in normally not available unless you just happen to know.
Professional Society Body of Knowledge	• There are trade magazines and professional societies for almost everything. • The societies are always looking for publishable information.	• Information is normally restricted to members. • Often a fee structure to provide funds to the group.
User Groups	Good source of free flowing information.	• Proliferation of "home remedy" and "folk story" class of information. • Usually little scientific discipline behind data and information. • Can become a life unto itself and take up a lot of time.

While you are performing your Technology Benchmark you may find something that is very close to your idea. This means that you are on the right track, just someone beat you to it. That doesn't mean you quit. If you persist, you can often find a spin-off idea or a twist to your original idea that may still have market value and not infringe on the prior art.

You can actually perform your own formal patent search if you like. However, I would suggest that you hire a professional at the point of preparing your patent application.

Exercise 4-6: Technology Benchmark

The Technology Benchmark is an extremely valuable tool once you master it. Think of yourself as an all time great detective – a modern day Sherlock Holmes - with your mission being to discover any products or research relevant to your idea. If you get really good at this, you'll find that you become an information sponge absorbing bits and pieces from your personal contact network, public library, newspapers, web searches, and periodicals.

Record all of your findings. That means take pictures, save web links, make folders and save any information you find in your design journal.

Make a list in your design journal and add to it, include as many key words and key word groups as you can think of. This is a good brain storming activity with your team. These key words and groups are what you will use in your searching.

Don't be overwhelmed by this exercise. This becomes almost a lifestyle after a while. I find that I am always collecting and cataloging information that might be useful on a future project that I am thinking about, rather like a packrat of information.

The Practical Inventor

Intellectual Property

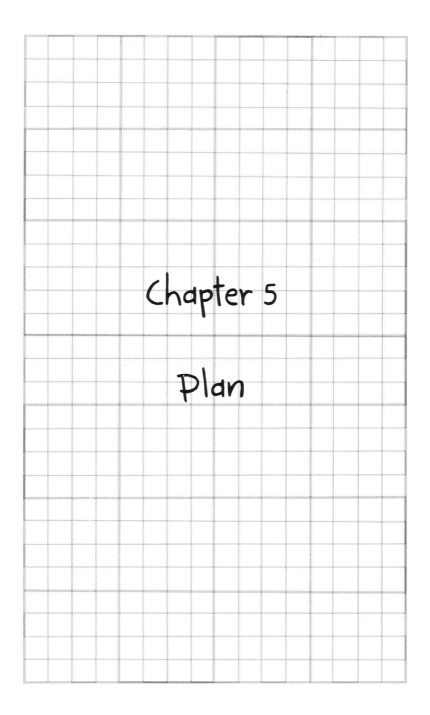

The Product Plan

It is time again to take off your inventor's hat and this time put on a business hat. Don't worry, this won't be so painful. It might even be a little fun because you are going to create a path to follow. In this chapter we are going to put together a product development plan. This is not a business plan or a plan for developing your business. We will, however, talk a bit about synchronizing your product development plan with the business development. When you arrive at the point of business plan development and creating your business development plan, there are a variety of resources available. We will talk about those in appendix A at the end of this manual.

People have different ideas about what planning is. For our purposes, we are determining and recording the steps to move your idea from where it is now to a point where it is a salable product. Start with you determining your goal for the idea and then move backward putting the milestones and steps into place to create a roadmap for your project to get from where you are now to your goal.

What caused you to purchase this manual? Was it to learn how to develop your idea? build a prototype? get a patent? sell it, make millions and retire to a beach in Belize and drink beer like in the TV commercials? I can help with a portion of that.

This chapter is important because you have an idea and you think that you can make some money with it. The plan is important because you probably have limited resources and a wrong turn along the way can be a major setback. The plan will not guarantee that you don't make any wrong turn or do the same thing over again but it can and will certainly reduce the number of times this happens.

If you are having thoughts right now about running out and getting a patent or licensing your idea to someone and your plan is based on one of these two scenarios, your product development plan will be slightly different. Don't worry about that right now, we will discuss the differences later. Right now, you are getting ready to roll up your sleeves and get to work on your product plan. The process you are going to follow is adapted from Randolph Craft's "Seven Steps of Effective Product Plans". Mr. Craft, of Pacific Planning Institute, Inc (see Appendix A), is one of the most gifted project planners I've had the pleasure of meeting. If you are at the point of extensive planning to launch your company, he would be a good resource. And as a bonus, you get the opportunity of writing off a trip to Hawaii as a business expense.

For a moment, let's revisit exercises 2 and 3 from chapter 2 on ideation where you drew out an initial project plan. At that step, you really couldn't go wrong because it was more of an exercise in identifying what you saw at the time to be the important steps in moving your idea forward. That was good and it was also more than most idea people ever do. What you are going to do now is fine tune and improve on it. This may change the order a bit and may sprinkle in other items along with detail. It was good to put your thinking onto paper and let it rattle around in your unconscious for a while? It's time to look at it again so grab your journal, open it to those pages and let's go.

Exercise 5-1: Product Plan Preparation

Creating a solid plan is a process not an event. Be prepared to take some time. You are going to work on this for a bit (could be several hours to several days for each point), set it down, come back and work on it some more, and so on until it is finished. I strongly suggest doing this exercise on your computer. If you don't have graphics software, just note the graphics and draw it free hand in your journal.

Scope – Your project scope is going to establish how far reaching your project is. The scope could be that your invention would affect every car on the road or only pick up trucks or your product would improve the efficiency of all vehicle gas ranges.

1. What is your project's scope? How broad and far reaching is it? Be realistic and as thorough as you can.
2. What is the importance or what need in society does your project address?
3. What is your target customer, audience, or recipient?
4. Identify Stakeholders – Stakeholders are people or groups with a vested interest in your success. They could be investors, friends, or immediate family members. A stakeholder normally does not have an active part in your project.

5. Who are the people who will be affected by the endeavor and who can influence it without being directly involved with doing the work? Families are stakholders, especially spouses and children.
6. Who are individuals or groups with an interest in your project's success in delivering intended results and in maintaining the viability of the project's product and/or service? Stakeholders influence programs, products, and services.
7. Who are the people with a vested interest in your project?
8. What are the explicit (very clear and detailed) expectations about the quality of your project?
9. Are there explicit expectations based on industry or government standards and regulations?
10. What implicit (implied or assumed) expectations could affect your project?
11. What can you do to turn implicit expectations into explicit expectations?
12. Constraints – those items that can throttle or gate your progress. List the constraints you can identify for your project.
13. Prioritize the list of constraints by what you feel is important.
14. Other than funding, what is your largest obstacle for succeding on this project?

Exercise 5-2: Project Goals

Setting goals – Goals are definitive milestones that are measurable and acheivable. As you go through this exercise be evaluate your goals to verify that your statements are measurable and acheivable.

Putting a plan together

If you've gotten through the previous exercises and not just skipped ahead, good job! You are going to use this information while putting your plan together. Your plan will guide your steps as you move forward. You will refer to this plan every morning as you have your first cup of coffee to plan your day. Putting it together is also going to take a fair amount of time but it is time well spent. One final thought on planning; it's never really finished. You should be constantly reviewing and tweaking your plan.

The usual project management and planning tools are the Gantt chart and the PERT chart. Those you talk to prefer one over the other but in reality, both are good tools. Since this is not an extensive manual on using these charts, the planning in this chapter is going to cover only the basics. You will likely want to learn more about the Gantt and PERT charts or hire a professional when you are ready to move forward. Don't forget: *all inventors need professional help.*

There will be a lot more that goes into the plan later. For example; looking at the sequence and how the plan you are laying out here fits in with your business development sequence. Your business development sequence is beyond the scope of this manual, but is critically important for your overall success. What we want to do here is provide the points where you should be synchronizing with your business development sequence. A lot more detail and information that will go into the project plan, you are just getting started.

The Practical Inventor

Plan

1. Start by writing the final outcome of your project on the first card and place it on the floor on the right-most part of your workspace.

2. Working backward, write each milestone needed to get to the final outcome on separate cards and put them to the left of the last card. There is likely more than one milestone. Don't be afraid to just blast out the milestones.

3. Continue working backward with your milestones until you get to the beginning.

4. More than one milestone can occur at the same time. Put them next to each other.

5. Some milestones need other milestones to happen first. These are called predecessors. Put the predecessor milestones in-line preceding that card needed them.

Plan

6. Remember talking about risks? I forgot about the cat. Maybe Randy's Post-its® are a better idea. ☺

7. Now start at the beginning (far left card) and work your way through the plan. Do these milestones make sense? Feel free to add, to remove and rearrange cards as needed to get the major flow of your process where it appears reasonable and doable.

Exercise 5-3: Framing the plan

Earlier in the chapter we talked about Milestones. Now it is time to identify and organize the milestones for your project. Once again, there many different tools you can use for this step including fancy computer graphics programs (which I love using) but we are going with a lower tech solution that works well at this point. I use 3-by-5 cards on the floor but Post-it® notes on the wall work just as well.

Following the guidelines presented in the above text, construct your initial plan.

Good work, you have just completed the basic framework for your project PERT chart. These are the major accomplishments you have identified to achieve your final goal for this project. Take a picture or copy your chart into your design journal before your cat saunters through. If possible, keep your chart to use on a later exercise. At least keep your cards or Post-its®.

Plan Evaluation

Let's take a look at the example:

First, notice the 3 main branches in the example chart. One of these branches will have the path that will take the longest time to complete. This

The Practical Inventor   Plan

will be called your **critical path**, and the one that you will have to manage the closest to make your time schedule.

Let's say the top path is your product development path and the lower path is your business development path. What is missing? *Links between the two main paths that establish the dependencies on each other.*

Projects vary so much from one to another that I cannot give you definitive linkage points, only some general rules and principles to follow. The important point is to identify the link points, understand them, and carefully apply them. Another variable in synchronizing your product and business development is the timing for raising funds. For instance, you as the inventor can build a proof of concept prototype but on occasion this requires outside funding. A good example of this variable is one client who built her proof of concept prototype from the kitchen trashcan and some duct tape while another client needed $64,000 for his proof of concept. Obviously, the need to raise funds will come in at different points. The following chart provides a plan that is a good rough outline.

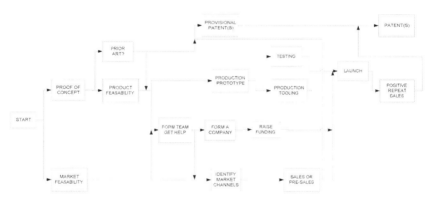

As you look at this chart, you will note it is pretty general. That doesn't matter, try to get an idea of the concepts and adjust it to fit your project rather than trying to fit your project into it.

Plan   The
                                                                      Practical Inventor

START – a good place to begin.

PROOF OF CONCEPT – Most inventors first build a proof a concept prototype or prove their concept through some preliminary research. If you are at this point with your project, read chapter 4 on Intellectual Property before doing any more with your prototype.

PRODUCT FEASABILITY – This would normally be performed in the process of evaluating the proof of concept prototype but I suggest doing this step first if possible. Why build a proof of concept if the idea is not going to make sense from a product or technology perspective?

PRIOR ART – This is an informal search performed by the inventor to discover and document related products or services in the pubic and private sector. It is not specifically a patent search. You should be looking in the public domain for any evidence of someone having your idea, which may or may not have a patent. While performing this step, you may discover new possible features and societal problems that could be covered by your idea or an expansion of it. This step is usually performed after the proof of concept unless the proof of concept would be expensive or difficult. You should do it at the same time as the product feasibility and market feasibility because much of the fieldwork overlaps. Depending of the size and cost of the project based on your idea, you may want to have a formal patent search performed for you earlier on, especially if you are feeling that your ability to perform this is limited. You need to weigh that cost against the cost and risk to your project. Chapter 4 on Intellectual Properties provides many tools for prior art and informal patent searches.

MARKET FEASABILITY - Look at what is being sold. Face it; your idea is worthless if you can't sell it, so do some shopping. Go to the library for sales data. This step should be performed early, even before the proof of concept. Most inventors build their proof of concept before they do anything so if that is your case, it should happen right after that. Keep your eyes open for related products when you have an idea. It's amazing how many people drive Fords all of a sudden when you buy a Ford.

FORM TEAM, GET HELP – Once you have evaluated your idea for feasibility and checked out prior art, you should form your team. The faster you do that, the more efficiently your project will move forward. (It took me years to learn this, I hope you learn this quicker.) Even though you may be capable of doing everything, you are likely not the best at each thing or even like doing it all. Why not put a team together of people who love

89

doing their part? (It might be difficult to imagine someone actually likes accounting.) You can look in Appendix A for resources such as CEO Space, CEO Mixer, SCORE, SBA, and SBDCs which are good sources of safe and knowledgeable help.

FORM A COMPANY – Now that you have a product idea and a team, form your company. But keep it lean. You don't need a slick office, employees, or lots of supplies. In Colorado, it is $25 with the Secretary of State on-line to register your company. After doing that, you need a 3 ring binder to get started. Worry about the rest when you get there. Right now, your resources need to be focused on moving your company forward and raising the funding to pay for patents, prototypes, production launch, marketing and sales. CEO Space is a great resource to help you move the business organization forward. If you are licensing, you don't need the production launch, marketing and sales but there are costs associated with presenting your product for licensing. Have you paid for a patent lately? A simple patent will cost you $5000 and the costs go up from there.

IDENTIFY MARKET CHANNELS – This step presents a great opportunity for some of your team members to pitch in and help. Decide what markets you are going to for sales and when you will hit that market. This will eventually turn into a marketing plan but for now, this step will provide the information you need for a complete business plan, raise the funding, and launch your product or license it. Do this step any time after forming your team so that you have help. Your team will usually want to be compensated for their time and most teams are compensated through founder's shares in the company, which have significant tax advantages. That is one reason why you should form your team before forming the company if possible.

SALES OR PRE-SALES – We are getting close to raising capital. You have put together a team, formed a company, and defined your product. Once you identify your market channels, create some sales literature and take a deposit on orders for delivery in a few months. This is a great way to generate some initial funding, just be certain you can deliver on the order. *If you use this pre-sales route, you should have your provisional patent filed with the patent office.*

RAISE FUNDING – Take note, you do not need patents or prototypes yet. Your initial funding is kept in private with signed NDAs with each prospective fund provider. If a potential investor does not want to sign the NDA, find someone else. After forming your company, raise funding (not money) so that you can set aside a salary for yourself for a year

Plan   The
                                               Practical Inventor

and execute your operating budget which includes funds for prototypes and
patents. A commonly used term today is becoming <u>UNSTOPPABLE.</u>
Raise your funding but be frugal once you have it. (Look in the appendix
for more information on funding.) This is still other people's money that
has been entrusted to you. Remember, this is you're your initial round of
funding; these funds are the most expensive for you at this early stage. If
you need to hire people and pay your team, you will do so after you've built
some value into the company. There are other routes for funding and while
it is outside the scope of this manual, there are funding resources listed in
the appendix.

PRODUCTION PROTOTYPE

– Now it is time to build a production prototype or a pre-production unit. Even if you are going for licensing, this term sounds a lot better than prototype. You've already built one prototype; you can make tweaks as needed on the pre-production unit. At this point, you may even be starting into the tooling because it makes sense to start building some of the lower cost production tools now too. The point is, the faster you get to production, the sooner you start creating revenue and everyone is happy. Now you and your team, need to define the features in your first product release. Remember that while you really want it so, your first product release does not need to be perfect, just appropriate.

TESTING

– Start testing your pre-production unit. The first testing will be to verify the soundness of the design so that you can release the production tooling while the testing goes on. To complete this milestone, you may produce more pre-production units and perform beta testing with your target customer for extensive field-testing in real world situations. You will want your provisional patent filed before any testing takes place in a public venue.

PRODUCTION TOOLING

– Once you have built your pre-production unit and started testing it to the point where you have confidence that the design is sound, launch the production tooling. Don't worry about it being perfect, just be diligent, have good thorough design reviews and release your product design. There will be changes and corrections. That is the nature of this business.

PROVISIONAL PATENT(S)

– Did you ever play the board game Risk? The key to success is strategy with certain things executed properly and at the right time. Patents are like that too. Chapter 4 on Intellectual Properties provides extensive detail on this discussion but for now; think about public disclosure and timing. If you keep your idea in the private sector, there is no a need for a patent. File your provisional patent

The Practical Inventor — Plan

(for as little as $100) before any public disclosure and as close to your market release as possible. The provisional patent is just a placeholder that protects your ability to file for a full patent for up to a year from that point. The provisional patent buys you time to assess the market viability of your product in sales before spending the money on a full and final patent.

LAUNCH – This could be replaced with licensing if that's your plan.

Depending on circumstances, several of the previous milestones may not be necessary for your licensing. However, every milestone you hit will potentially increase the value of your product and the size of the licensing deal. Testing may or may not be completed before launch but you have enough test results to have confidence that your product is ready to go. By now you want the provisional patent(s) on file and ideally some product pre-sold.

POSITIVE REPEAT SALES - Repeat sales are key because

many people will buy something to try it but only the ones who like the product will come back for more. This is one of those key indicators that tell you that you've done it. Yes, the first sale is exciting but the first repeat sale is the true measure of a successful launch.

PATENT(S) - If your sales are good, you have a strong market with

good profits, and repeat sales, you may still decide to forgo the patent. The chapter on Intellectual Property addresses this further. Just be sure it makes sense from a business perspective. When you reach this point, drop me a note and some positive feedback on how this manual helped you get there. Oh, and good work!

Wow, that's a lot. Now go back to the chart you made and look at it again. Feel free to make adjustments as you see fit. That is OK; even the best and most experienced will rearrange things after leaving it for a bit and looking their plan over again. Are you ready to keep going? Good!

Remember the chart you've created containing the milestones is not your plan. This is the framework of you project plan. Take a step back for a moment and think about moving your idea forward. Take time to organize your thoughts and get some clear direction! Compare it to building a house. You have framed and going to close it in and finish it.

Plan

At this point you can swap the next two exercises depending on your constraints and preferences. Do it however it works best for you.

Exercise 5-4: What Happens Next?
1. Get out your calendar.
2. Start on your right, the final outcome, whether it is license, sales, or launch and write your target date.
3. Now work backwards from end to beginning and put down target dates for hitting each milestone.
4. Once you've gotten back to the start, go back through each path.
5. Do your best to evaluate if the time between milestones is reasonable.
6. Feel free to add milestones any time that they come to mind.
7. Adjust your milestone times as needed.

Progress Check

You might be feeling like you've spent an eternity in this chapter when you really want to be out building something. I feel your pain. We are almost done. Look at it this way, instead of creating a product from your idea, you are creating the process to create a product and make money from your idea. It is a proven fact that taking the time up front to plan and have an organized, methodical approach will save you a significant amount of time and money.

You now have the process and milestones mapped out in this PERT chart along with target dates so you have an actual date, on paper, that shows when you can start seeing a pay check and when your new company will be seeing a positive revenue that will lead to profits. That has got to get the blood flowing in those veins a little faster! The million-dollar question is "How are you going to get from one milestone to the next one?"

The Practical Inventor   Plan
Exercise 5-5: One Step at a Time

On some projects, you will want to translate what I am talking about into the computer or you'd end up needing to rent the high school gymnasium. If you were not able to keep your card chart from exercise 1 intact, now would be a good time to rebuild it.

1. Take a different color, colored ink, or different size card for this step. You just need to be able to visually distinguish between your milestones and intermediary steps.
2. Begin at the START. I'd say start at the beginning but that is not what the box on the chart is called. Identify and write on a card each step or task you can see that needs to be completed to get from the START to each of your first Milestones.
3. Repeat this until you get to the last Milestone on your chart.
4. Go back over and make your best guess for the time needed to complete each task or step.
5. Now go drink a beer or two and relax a bit.
6. Look over your chart and re-arrange as you see fit.
7. Finally go through each task and list the resources you will need and a guess at the cost for that task.
8. Document your chart with pictures and probably several sheets of graph paper.
9. I would recommend that at this point, you obtain a good project management software package such as Microsoft Project® although it is better with Gantt charts it does provide for putting in your timing, resources, and cost estimates and then track your progress. The PERT chart presentation is not as effective as some other software that is available.
10. I would also recommend at any point in this process you feel overwhelmed, take a break, have a beer, or whatever you would do to relax and re-evaluate. If you are still overwhelmed consider hiring some professional help.

Chapter 6

Prototype

Prototype   The
                                        Practical Inventor

Prototype

Finally we get to the chapter on building prototypes. I don't think there is anything as rewarding as seeing an idea that has been rattling around inside this thing between my shoulders become a reality. The reality is that anyone can build a prototype. The question to answer is how do you build a prototype in a manner that moves your idea into a salable product without

The Practical Inventor — Prototype

spending your life savings and second mortgage to accomplish it? The answer to this question lies in having a plan and using some of the creativity you have that generated the idea to start (OK you might still spend your life savings but you'll get a lot farther).

The purpose for this chapter is to help you to understand the prototype process and to understand the different types of prototype technologies available along with the relative costs, strengths and weaknesses of each method.

One of your challenges is to reduce the number of prototypes you build between your first idea and the production launch. I've seen prototypes built to the tune of several thousands of dollars just to be tossed aside to build another because of some little thing that could have been avoided. I've also seen as much as $60,000 to build a prototype when a glue and cardboard model built by the inventor would have been just as valuable for the stage and purpose for the prototype. A product that sells for $15 will typically cost $2 to $5 to manufacture but $500 to $1000 for a prototype. You can see the obvious advantages of reducing the numbers of prototypes. When you know that you will be building a number of the same vintage prototype, there are intermediate methods for reducing the prototype cost that is covered in the section on building prototypes.

As you may have guessed by now, this chapter is dealing with planning and sequencing in addition to building your prototype.

Prototype   The
                                       Practical Inventor

The different reasons for building a prototype:

There are many reasons for building a prototype. In general, the more finished and production representative the prototype, the more it will cost. Most of the prototypes built do not need to have the fit and finish of a production part. In fact, you would be surprised how far a cardboard and duct tape model can take you. I used a model built out of foam core board and packing tape along with some parts from my "junk box" to raise $275,000 in development money because it was enough to show that the technology could work and also showed my investor that he would be investing in someone who understands how to use his money properly (the new Corvette comes later).

The main reasons for prototypes are:

a) **Proving a concept** – This is often called a proof of concept. "I just had this idea and wanted to see if it would work." Been there, done that? Me too.

These prototypes are often made with cardboard, foam core, duct tape, and glue. It could also be a bunch of wood, metal, or other material pieced together. Sometimes, the proof of concept is research based because it would be too costly even a very basic model, or not possible.

If the invention is electronic, it could be a breadboard circuit or some type of point-to-point wired circuit. In any case, the proof of concept is not pretty and often has very limited functionality but sufficient to prove the concept can work.

**Electronic sample
Courtesy of Veritek, LLC**

One inventor I know salvaged a bowl from her kitchen trash and with a few extra pieces, and some duct tape, has a proof of concept.

99

b) Demonstrate a principle

This is Proof of Concept of a part of an invention. When a product is too complicated or too expensive just prototype the portion needed to demonstrate the part of the idea that you need to prove.

c) Testing and product development

These are also known as engineering prototypes. These prototypes usually have a few versions as the product development moves forward and testing results require changes. In general, the more complicated the product, the more versions of prototype. It is common to build several test units before going to production. It is also common to use the pilot production units for this testing to reduce the number of prototypes that must be built.

d) Show your idea to solicit help or funding

Sometimes you need a prototype to demonstrate your idea to someone. This may be needed to convince people to join your team or to raise working capital to move forward. The one certainty is that you cannot use the proof of concept. It is possible however to use the virtual prototype or a more finished prototype for this if needed. There is almost always a prototype you've built for some other reason that you can use here. The level of finish really depends on the level of help or funding you are looking for. Obviously a higher level of finish is needed when raising $10 million than when raising $10 thousand.

e) Show for marketing or sales:

This might be a product for a trade show. This part will usually need to look like a production part and it usually needs to work.

f) Demonstrate a finished product

Demonstrating a finished product implies having a finished product or at least, the appearance of a finished product. The most often reason for this is to have product to show at a trade show. More products shown at the major trade shows than anyone will ever admit are "smoke and mirrors". I've seen and had my fair share in several.

g) Pilot Production (pre-production)

While the Pilot production models are not really prototypes any more, it is worth noting so you see where things fit in the overall process. Pilot production is ideally performed with production tools. Sometimes however, it is good to have a pilot production run using prototype parts. This is normally the case when the tooling to make a part is extremely expensive or very long lead time and you want to be absolutely sure of the final design before moving forward.

Different types of prototype technologies

We started talking a bit about different types of prototypes in the previous section when we were talking about the different reasons for building a prototype. It is to your advantage to select the appropriate prototype for its purpose. Failure to do this often results in spending much more money on prototypes than needed and takes that badly needed money from someplace else often stalling or stopping a project altogether.

Let's take a look at the different types of prototypes:

a) Proof of Concept

The proof of concept is the least expensive to build. These prototypes are usually not the type of thing you want to show other people because they can be pretty crude at times. Further, if you don't know for sure that it really is going to work, why spend any more money on it until you can prove it out, or at least part of it.

b) Virtual prototype

With the technology that we have available on the computer, you can design and simulate an electronic circuit, or design a mechanical assembly, test the components for strength, mold filling for a plastic molded part, and the operation of the assembly all before building a single part. As you move closer to building other types of prototypes that we will cover later in this chapter, it will become necessary to create the 3D models of the mechanical parts so the mechanical assemblies can be tested for very little additional cost and could save a lot of money rather than building the prototype over again because the first one didn't work.

The same is true for the electronic designs. The key is to select the appropriate CAD system to enable the testing. A fully functional

CAD system with these capabilities is available for under $6000 in 2006 prices. Then you have a learning curve so this is an opportunity to look at the "getting help" part. In chapter 7 on Launch we discuss good practices for hiring this help.

c) Engineering prototype

Engineering prototypes will vary in the fit and finish from being not much better than proofs of concept to looking like a finished production unit and everything in between. In fact, for many projects, your engineering prototypes will progress through this sequence called product development where it becomes more refined as your go. Hopefully, your first engineering prototype will have the benefit of being tested as a virtual prototype first. It is possible that our first engineering prototype could be your only engineering prototype which saves you a boat load of money and time. The main reason for building an engineering prototype is to test your actual product as it would appear in its final form or to test a particular piece of your idea but may not have the final fit and finish of a completed product to save a little on your budget. It is also not uncommon to build multiple copies of your engineering prototype for extensive testing, life testing, and beta testing with end users.

d) Marketing prototype

Marketing prototypes are built primarily to sell your product or to raise capital. This may be for licensing to a prospective interested party, display at a trade show, or to sell your product on a home shopping channel. I am guessing that you are thinking of many other potential uses for it and you would be correct. These are usually the most expensive prototypes to build because they need to look, feel, and operate like the final production unit. I have on many occasions, taken an engineering prototype and finished it to look and feel like a production unit.

Wow! Think about this. After testing your virtual prototype, build one engineering prototype, then polish it, paint it, and use it to show at your trade show booth. Think it can't be done? Guess again, I've done this many times. In fact, I'm doing it right this moment with a product that I am showing at a trade show in a few weeks and will be using this same prototype soon to get onto a home shopping channel for sales. It comes down to good planning and excellent execution of your plan.

Exercise 6-1: Your Prototype Strategy

Often, what type of prototype you build depends on your stage of development and time table. If you are just in the ideation stages and there is a trade show opportunity coming up that is a once a year gig, do you wait until next year or pull something together? Unfortunately, sometimes we have to wait for next year but we won't thing about that right now.

Revisit your product development plan and add what prototypes you will need at the appropriate points in the process.

Prototype Technologies

Just like having many choices for travel from Denver to Dallas, there are many choices for building a prototype. The different available technologies have their respective pros and cons along with the associated costs. Just like with traveling you choose different modes of travel based on different characteristics, you choose the prototype method normally based on a combination of the purpose and your available budget. The prototype methods presented in this chapter are not every possible method but cover a good range.

The various technologies of prototypes we will discuss are:

- Cardboard and glue
- Clays
- Machining
- Injection molding
- Vacuum forming
- Stereo lithography
- Stereo laser sintering
- Fused deposited modeling
- Casting
- 3D printing
- Electrical bread boarding
- Proto-Board
- Quick Turn boards
- Virtual prototypes

Cardboard and glue

Positive:
- Very quick and inexpensive
- Does not require drawings, only imagination.
- A trip to the local craft supply or hardware is normally all that is needed.

Negative:
- Not very pretty
- Fragile
- Not something to show others

Typical uses:
- Proof of concept

These prototypes can be built using plywood, foam core, carved blocks of foam, duct tape, salvaged items, paper Mache, or even paper models. This is the most basic form of a prototype. It is often the method used for proof of concept models and rarely leaves the inventors' shop. Part of that is because any attempt to move it can result in disaster. All jokes aside, these are pretty crude models that work well to try out a function or to simply prove the concept.

This category prototype can be generated from napkin sketches, 3D CAD drawings, paper drawings, or straight out of your head. The inventor is usually the one who builds these models as there is often little or no documentation yet and this is an attempt to get a "feel" for what the product could be as a result of the idea or invention.

The Practical Inventor

Prototype

The pictures shown below are foam core and glue, working models that took about 3 hours each to make from CAD generated drawings. It is common to just sketch out the idea onto cardboard or foam core and cut out the part directly. I've also made plenty from napkin sketches.

 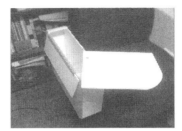

Samples courtesy of Veritek, LLC.

As you can see in the photos, they are not pretty. The hinge on the right side is packing tape. The edges are sharp, square edges rather than nice rounded edges, and they are quite fragile. They did prove the concept nicely.

This picture is made from duct tape, a salvaged bowl and a used water bottle with a lid. An inventor friend of mine built this but then decided that it was not worth pursuing after finding several similar units already in the stores.

Sample courtesy of Lisa Barker

Prototype   The Practical Inventor

Clays

Positive:
- Allows for a very interactive and creative design process.
- Normally, only sketches and imagination are required.
- Allows for full size prototype at a very early stage for deign and styling.

Negative:
- Very time consuming and labor intensive.
- Fragile heavy models.
- Non-functional, usually for styling only.

Typical Uses:
- Styling design
- Proof of concept

Clay models are an old art that is still being used. In this technology, a model is sculpted out of modeling clay. This may sound like a throw back to first grade but some of the models are highly detailed. Even today, automotive design studios do full size clay models of a new body style because it is something that you can walk around and get a sense for what the real thing will look like. As you can imagine, a clay is very time consuming and expensive, mostly labor. The clay models happen very early in the design and styling phase and are normally generated directly from the head of the designer aided by some sketches.

Machining

Positive:
- Normally no tooling needed only a CAD file and nominal setup charge.
- Can be a quick turn around on the parts.
- Can be machined from a sketch but CAD model is better.
- Prototype parts can be fabricated from actual design materials.

Negative:
- Can be more expensive than other methods.
- Can take longer than other methods.

Typical uses:
- Proof of concept
- Engineering prototype
- Pattern for cast prototype parts
- Marketing prototype
- Pilot production parts
- Production parts

Machining parts is another older but very reliable method for building prototype parts. In some cases the prototype is the pre-production since the part is designed for machining anyway. As the name suggests, parts are created on a machine called a mill, CNC (stands for Computer Numeric Control), lathe, wire EDM (Electronic Discharge machine) or other types of machine tools. Molded parts are often created as machined parts for the prototype because extremely accurate parts can be cut from the actual target materials that will eventually injection molded. These parts can be machined from sketches, drawings, or 3D CAD files. Some suppliers machine the parts directly from your 3D CAD model and use whatever drawing you have just to check their work.

Machined parts also include short run sheet metal, laser cutting, and water jet cut parts.

Prototype   The
                                         Practical Inventor

The following pictures show examples of machined prototypes.

3D CAD model   Machined parts   Completed assembly

Parts could be laser   Aluminum separator tool
or water jet cut. This
part was water jet cut.

Images courtesy of Veritek, LLC.

Injection Molding

Positive:
- Production quality parts
- If the parts are good, you are ready for launch

Negative:
- Tooling cost is still high
- Longer lead time to produce the tooling

Typical Uses:
- Engineering prototypes
- Marketing prototypes
- Pilot production
- Production

Injection Molding is a production process where plastic resin is melted and injected into a metal cavity called a mold. These molds are normally made for steel and can last for millions of parts. However, there are prototype methods for injection molding parts where the tooling cost is much lower but the individual part cost is higher, usually 2 to 3 times what the production part would cost.

One common method for a low cost molding tool is to make it out of aluminum instead of steel. This cuts the cost because the toolmaker can cut the aluminum much faster than the steel. Eliminating cooling lines in the tool base and eliminating the ejectors can reduce the tool cost further. The result is that the individual parts cost more money because of the additional labor and extended cycle times. The softer aluminum tools also work against you because the same properties that allows the lower cost tool, also means that it wears out faster. If you only want 50 or so parts, this is a good way to go.

Almost every injection molding shop has prototype parts programs. Some provide a small number of parts at very reasonable rates because they have developed low cost and fast methods of generating the prototype tools. This method usually requires the 3D CAD model to feed into their automated tool design and processing.

Prototype   The
                                           Practical Inventor

---

Vacuum forming

Positive:
- Production quality parts
- If the parts are good, you are ready for launch

Negative:
- Tooling cost is still high
- Longer lead time to produce the tooling

Typical uses:
- Engineering prototypes
- Marketing prototypes
- Pilot production
- Production

Vacuum forming is also a production process that has a prototype alternative. Vacuum forming uses a sheet of plastic that is heated until it is really soft, then it is set onto the mold and a vacuum is channeled through many pin sized holes to suck the soft plastic onto the mold. The mold is cooled and the part is removed. Like the injection molding, the prototype parts are fabricated on the same machines that your production parts would come from except the tools for forming the plastic are in this case made from wood or REN which is a cellulose (like wood) block that is readily machineable into a prototype tool that can also produce somewhere on the order of 50 prototype parts.

The
Practical Inventor   Prototype

Stereo lithography (SLA)

Positives:
- Generally considered the most accurate RP process, and the best surface finish
- Accuracy: +/- .005 on a 5 inch part, +/- .0015 inch per inch on parts > 5 inches
- Wide range of photopolymers available
- Ideal as patterns for an RTV mold for making cast parts

Negatives:
- Models have to be sanded for better appearance
- Material properties degrade over time / moisture absorption

Typical uses:
- Engineering prototypes
- Marketing prototypes

Stereo lithography (SLA) is in a category known as rapid prototyping because it requires no tooling to fabricate the parts. This process creates solid models by curing a liquid polymer resin with a laser beam usually in .006" layers with a .010" diameter laser beam.

It is the most widely used prototype technology and generally considered the most accurate process. Making parts with the SLA requires a 3D CAD model. The computer slices the CAD model into layers .006" thick and draws each layer slice in the tank filled

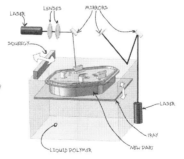

with resin that looks like a deep fryer vat filled with hot grease. A tray in the tank moves up and down. Making the parts is called "growing" the part because the tray starts .006" from the top. A squeegee like device ensures an even layer of resin and the laser does its thing. The tray moves down a layer thickness, squeegee, laser and so on. In that manner your can actually watch the parts grow in the tank.

Prototype
The Practical Inventor

It is not quite as bad as watching grass grow but several people would probably argue that point with me after watching the process a few times.

Image courtesy of Protogenic, Inc.

There are five finish levels available for SLA parts. These levels are:
1. Raw SLA – The part (bottom left in picture) is not finished, the .006" layers are sharp, clear, and pronounced. This method allows you to save money by putting the level of finish you want on it yourself.
2. Functional – The part (bottom center in photo) is finished to remove the clear lines of the layers and provide dimensional accuracy. This is the most common level of finish that is ordered. This is also the level of finish when using the SLA for a pattern to make a silicone mold.
3. Clear Coat – A clear coat (bottom right in photo) can be applied to seal the resin from absorbing water. Some prototype shops will finish the part a bit further before applying the clear coat.
4. Polished – The part (top right in photo) is polished to the point where it can be optically clear. With some of the resins available, you can make colored lenses and clear windows.
5. Paint Ready – These part (top left in photo) is finished more than with the clear coat parts but not generally polished to the level of the clear parts of the polished level and a primer coat is applied, ready for painting.

Some of the resins that are available for SLA parts are:
7580 and 5260: Durable, "ABS-like" resin, can also be post-cured for higher heat deflection (>98°C).
15120 Nanoform: Ceramic, high stiffness, high heat, very low moisture absorption.
5190: Accurate, also more durable than other rigid resins.
5240: Durable, flexible resin, polypropylene like.
Water Clear: (Somos 10110 and 10120) Great overall resin, can be used for light pipes and lenses.

Stereo Laser Sintering (SLS)

Positives:
- Parts are made with real thermoplastic materials, metals and elastomers.
- Parts can be layered and stacked within the build envelope to reduce cost.
- Parts are suitable for Direct Manufacturing, high HDT, chemical resistance, and are biocompatible.
- Complete functioning assemblies can be grown pre-assembled.
- Prototype parts can be fully functional.

Negatives:
- Relatively poor cosmetics.
- The surface finish will resemble a very heavy pebble texture.
- Accuracy / repeatability worse than SLA (+/- .3%); fine features a problem (min. laser line width is .026").
- Not suitable as a pattern for silicone molds.
- Parts are not watertight without filler added afterward.

Typical Uses:
- Engineering prototypes
- Marketing prototypes
- Pilot production

Stereo Laser Sintering (SLS) is a process similar to SLA except it uses very fine solid powder instead of the liquid resin. The SLS process also uses the 3D CAD model and slices the part into .006" layers for the fabrication.

One very unique feature of this process is that complete assemblies containing moving parts can be grown as a finished assembly rather than parts that must still be assembled. The carabineer clip (pictured later) is one example that even incorporates coil springs in the assembly. Parts or assemblies can also be fabricated in like fashion using stainless steel and a rubber-like material.

Prototype

The Practical Inventor

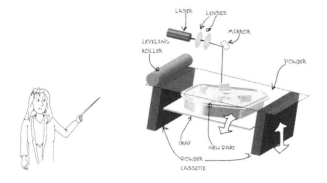

Powder is laid out with a roller, and a laser sinters (heats and fuses together) selected areas causing the particles to melt and then solidify.

The surrounding powder acts as the support structures, then falls away as the part is removed from the machine.

The laser beam is .023" diameter, and .006" layer is typical so the resolution for growing the part is .023" horizontally and .006" vertically. The finished surface very closely resembles the surface of a cast iron engine block.

The powder can be Nylon, glass-filled nylon (glass beads), metal (stainless steel is popular) or rubber like elastomers.

The sample part (provided courtesy of Protogenic, Inc.) is a production clip assembly. It contains 5 moving parts and two springs. The assembly comes out of the SLS machine complete with no post assembly work.

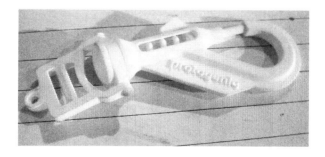

The Practical Inventor   Prototype

Fused Deposited Modeling (FDM):

Positives:
- Accuracy improved, on par with SLA's.
- Uses real thermoplastic materials (ABS, PC, PS, PPS).
- Requires no finishing after supports are removed.

Negatives:
- Poor cosmetics due to the extruded bead. The finish can look like the part was made from "squishy" thread.
- Not suitable as a pattern for silicone molds.
- Bonding between extruded layers can be weak.

Typical Uses:
- Engineering prototypes
- Marketing prototypes
- Pilot production
- Production

Fused Deposition Modeling (FDM) is an automatic extrusion process. I like to think of it as plastic extrusion meets CNC. The computer takes your 3D CAD model and creates a path to follow and build your part. If you've ever tried filling a really big crack with a calk gun, you have an idea of how this process works.

FDM is an extrusion process where materials are forced through a heated nozzle, which moves around the platform, forming the layers similar in concept to creating an object using a tube of frosting, or glue gun. The layers are 005" or .010" thick.

Materials can be ABS, PC, PPS, PS depending on machine.

116

Casting and Rapid Prototype Molding

Positives:
- Multiple prototypes can be produced relatively quickly and inexpensively as compared with other prototype methods.
- Several material options can be tested for relatively low cost.

Negatives:
- The "first off" part is expensive as the mold must be made first.
- Accuracy is not quite as good as with an SLA.
- Requires an SLA as a pattern for the mold.
- Part fabrication can be labor intensive.
- Casting is as much an art as it is a science.

Typical Uses:
- Engineering prototypes
- Marketing prototypes
- Pilot production

Casting and Rapid Prototype Molding (RP Molding) are processes of casting or molding parts using silicone molds. An SLA is used as a pattern to create the silicone mold. This process is very useful when you want to make several prototypes of the same version for testing, pilot assembly, or to show different features such as color. This process can also be used to make prototype parts with integrated rubber seals.

Silicone boot
Sample courtesy of Protogenic, Inc.

The Practical Inventor

Prototype

Notes on the castings

- Different silicones are used.
- Gates and vents are put on non-cosmetic (hidden) areas whenever possible.
- Typical mold life is 25-35 castings, but it is very geometry and material dependent. Some molds have yielded greater than 60 pieces when the parts have simple geometry and use soft material.
- Epoxy molds are now available for some applications that will last 100's or parts.
- Often times your castings look better than your injection molded parts.
- There are 15 different materials available, with more in development. Soft Shore A, flexible, vinyl-like, rigid, water clear. Most end products can be approximated in urethane and cast silicone parts.
- Inserts such as Helicoils are available for installation into your model.
- The parts are paintable, printable, and ready to show.
- Once a mold is made, you can produce a number of cast parts and experiment with several materials, durometers and colors all in the same mold
- Two-part and component over molds can be fabricated in this process.

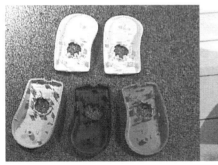

Different color parts
Image courtesy of Protogenic, Inc.

Elastomeric seals in part
Sample courtesy of Protogenic, Inc.

Prototype   The
                             Practical Inventor

3 D Printing

Positives:
- Part fabrication is very fast.
- Process uses low cost materials.
- Multiple colors can be used as the parts are grown in the printer.

Negatives:
- Poor accuracy, especially in Z.
- Parts have a poor surface finish.
- The materials have poor mechanical properties; require secondary infiltration to keep them from crumbling.

Typical uses:
- Proof of concept prototypes
- Early engineering prototypes

3D Printing uses a printing device similar that is a cross between an SLS machine and an inkjet printer. The most popular 3D printers use a very fine powder that is sprayed with a binding agent by a print head. Some use standard inkjet print heads and cartridges for the binder.

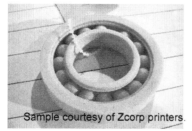
Sample courtesy of Zcorp printers.

What is cool about this process is that it is basically like a big inkjet printer in your office. Just start the "print job" in the evening and your part is finished before morning coffee.

Similar to laser sintering:

- A roller puts down each layer of powder.
- Inkjet head sprays binder, layers usually .010.
- Powder that has not been sprayed with binder forms the support structure.
- Material is cellulose or plaster

The
Practical Inventor   Prototype

Electrical Bread Boarding

Positives:
- Very fast
- Inexpensive

Negatives:
- Limited to slower circuits and low frequencies due to high capacitance in the breadboard itself.
- Not suitable to surface mount devices.

Typical uses:
- Proof of concept prototypes
- Early engineering prototypes

Electrical bread boarding is the process of building an electronic circuit onto a breadboard for test and development. The breadboard has rows that are connected and have spring-loaded fingers to insert components into and make electrical contact.

Sample courtesy of Veritek, LLC.

Prototype   The Practical Inventor

Electronic proto-board

Positives:
- Very fast
- Inexpensive
- Better bandwidth and high frequency performance than breadboard.

Negatives:
- Not as easy to change the circuit as with the breadboard.
- Not suitable to surface mount devices but possible.
- Not very aesthetically pleasing and sometimes a downright mess.

Typical uses:
- Proof of concept prototypes
- Early engineering prototypes

Electronic proto-board is another quick and inexpensive method of building a prototype electronic circuit. This is similar to bread boarding except the parts are actually soldered and hard-wired together. This method can use surface mounted parts but is a bit tricky. However, this is a good method to create a fully functional circuit to package in a rapid prototype part for a product test or demonstration.

Sample courtesy of Veritek, LLC.

The Practical Inventor
Prototype

Quick turn boards

Positives:
- Prototype can look and perform like a production board.
- Test final production design before committing to tooling. Especially important at high frequencies.

Negatives:
- Requires software to generate the files needed to fabricate the boards.
- More expensive than other methods.
- Takes more time to create the board layout and order the boards.

Typical uses:
- Engineering prototypes
- Marketing prototypes
- Pilot production

Quick turn boards are another electronic prototype method. Many if not all of the circuit board producers have small run prototype offers and there are a few companies that do nothing but prototype boards. It is also possible to make your own boards but I've found that with the programs offered by the circuit board producers and the environmental issues, I am not even temped to make my own any longer. If all you are doing is prototype, there are even on-line services that provide their own schematic capture and board layout software for free. The good side is that this software is expensive to purchase so this is a good deal. The down side is that you must use the people who provided the software to produce the boards.

Samples courtesy of Veritek, LLC

Prototype   The
                                             Practical Inventor

---

Virtual Prototype

Positive:
- Fast and relatively inexpensive method to prove a concept or design.
- Can save a lot of time and money over repeated physical prototypes.
- Allows the inventor to try variations.
- Can quickly optimize a part or assembly design.

Negative:
- Software is expensive and has significant learning curve.
- There is nothing like something you can hold in your hand.

Typical uses:
- Proof of concept prototype

Virtual Models are a fantastic method to prove out your design before committing to building a physical prototype. In the electronic realm, you can simulate circuits with models of real components. I love the circuit simulation because when the magic blue smoke comes out of the electronic parts, it is just a couple mouse clicks and the circuit is repaired and operating again. In the mechanical realm, the actual materials are modeled and you can obtain overall mass, center of gravity, strength of a part, find out when something will break, see how a plastic molded part will fill the cavity and correct problems before they happen. It is even possible to operate (and animate) mechanical assemblies to check for clearances, interference points, and proper operation. Obviously it takes some pretty expensive software, several thousand dollars worth, or hire the work out but you will need the 3D CAD model, schematics, and board layouts anyway so the incremental cost is nothing compared to what the typical savings are.

Rapid Prototype final notes

One final note is that there is a special class of prototype referred to as "Rapid Prototypes". This class of prototypes includes SLA, SLS, FDM, and 3D printing. These are called rapid prototypes because once you have the 3D CAD model; a detailed and accurate prototype can be produced in a matter of minutes to hours depending on the volumetric size. In fact, the defining characteristic is that a Rapid Prototype is produced directly for a CAD file. The other methods of prototypes are the older methods to produce the prototype (such as machining) but each has one or more advantages and drawbacks over the others. Hopefully this information will guide you in making the right choice. Don't be misled to believe that because it is called Rapid Prototype, you will get your parts more rapidly than from some of the other methods. A great website with more information is http://home.att.net/~castleisland/techn.htm.

Planning your prototype strategy

Now that you have more knowledge and hopefully better understanding of the different types of prototypes, their uses, strengths and weaknesses, you will put together a strategy that outlines your steps to move your idea forward with the prototypes as needed. Let's put your new knowledge to work.

Exercise 6-2 Prototype Planning:
1. Refer back to your product plan from Chapter 5 and make changes accordingly based on what you have learned about prototypes.
2. Make a list of the prototypes that you have determined you will need in the **Prototype** column of the table.
3. Add for each prototype:
    a. When you expect that you will need each prototype.
    b. Purpose for prototype
    c. Technology or technology mix for each.
    d. Any final notes

Prototype   *The Practical Inventor*

Getting Help

There comes a time when we all realize that we cannot or should not do it all. Maybe you are not there yet? When you get there, it will be a good idea to understand some of the pitfalls of hiring someone to design or build a prototype for you.

Remember that getting professional help is not a sign of weakness: it is a sign of wisdom and cunning. Go to the guys who do the one thing such as 3D CAD design, or prototype fabrication to have your work done and you are leveraging all of their knowledge and experience for your project with no overhead costs. You are just paying for what you need.

Your process for getting help should be:

1. Find possible service providers. There are many sources of help regardless of what you are looking for. In Appendix A is a list of some service providers. Check www.thepracticalinventor.com for an updated listing. You can also just look in the phone book or search on-line. There may also be an inventor club in your area that could supply adequate assistance.
2. Qualify your prospective service provider. If you are going to spend some of your hard earned cash, let's make sure it is spent in a way that will move your product forward effectively You should look into the following for each potential service provider:
   a. Execute a Non Disclosure Agreement (NDA) with the prospective service provider.
   b. Refer to your prototype service provider to ensure the CAD system that your designer uses is compatible with their systems.
   c. Check with your prototype fabricator. They should know the designer you are considering. Ask the designer for references to prototype shops they use. Ask the prototype shops for references.
   d. Ask your service provider for a list of clients and projects they have completed in the last 12 months. Check randomly with a couple of their clients.
   e. Take a look at some of their work. If you have a signed NDA they should not have a problem with this. If they do, just find someone else.

3. Enter into a service contract. In many cases this is nothing more than a quote from the provider to you and a purchase order from you to them. At minimum these contracts should specify:

a. What are you buying? Be specific about this so there is no question when it is delivered that you are getting what you asked for.
b. A specific delivery date. When will the work be completed and delivered?
c. A fixed cost for the service to be provided. When I quote jobs in a Statement of Work, often the project is broken into several phases. The first phase is a discovery phase where we obtain a clear definition of the project. Each phase is quoted with a fixed cost for that phase.
d. The contract should also provide for some compensation to you if the service provider fails to fulfill the contract. My preference is for the contract to contain an on-time delivery bonus rather than a late fee. This keeps the relationship in a positive frame of mind and you get to celebrate successes rather than bemoan contract violations. Do this by taking what would be off the cost up front and add it back in as the on-time bonus.
e. If your prospective service provider cannot give you a fixed cost, specified delivery date and agree to specific deliverables, you are talking to the wrong people, find a different service provider.

Don't get wigged out when they ask for money up front. An advance is normal but under no circumstances should you ever pay the full price until the product is delivered to your satisfaction. With new clients, I often get 50% of my labor and all our of pocket expenses paid up front or I don't take the job. My Statement of Work would get signed by you and become the contract between us. The contract is enforceable in court so both parties are protected as much as is reasonable. If you are buying a machined part from a machine shop, that's just a purchase order and pay COD. If you are hiring an engineer to design a circuit board, you may want to consider working up a Statement of Work.

Taking Action

You can only plan and study so much! You are an inventor and need to invent something. That means it is time to take action and build your dream.

Prototype   The
                                        Practical Inventor

Proof of concept prototypes:

The best way I can thing of to give you some ideas on building your prototype is just show some examples.

Example 1: Build a foam core and tape model

This will show one of many methods for building a model. You can use a variety of materials from cardboard and glue, to plastic sheets and solvent.

1. Start by sketching out on the foam core your design for each piece of the model.

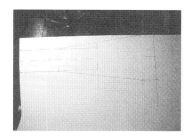

2. Now cut out the pieces with scissors or a sharp knife.

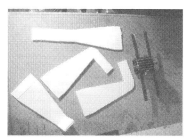

3. Tape the pieces together.

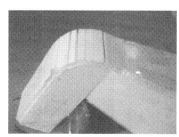

The Practical Inventor

Prototype

4. You can make bends by scoring the one side in strips and bend it around the radius.

5. Add other parts to it as needed.

6. If you want to dress it up a bit you can ht it with some spray paint. Go one step further by making a label, print it, and use a glue stick to attach it.

Prototype
The Practical Inventor

Example 2: Mounting a block on a box

This example demonstrates a fairly basic and easy way to machine parts:

1. Mark the part to be machined. In this case, I used a paper template created with my CAD system. You can also use a sketch drawn on a napkin or graph paper, or measure and draw right on the part like we did on the foam core.

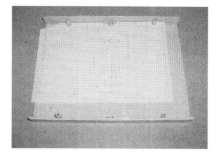

2. The "machining" might be really extravagant like an electric drill and utility knife are use a milling machine. Machine the part as needed, take your time and do your best. Remember, it probably does not have to be perfect, just good enough.

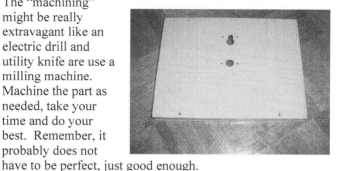

3. Mount the block using some machine screws.

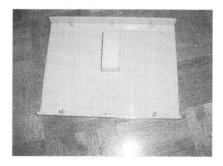

The Practical Inventor

Prototype

Example 3: Make a custom molded part

I needed a pair of flexible rubber hands for a prototype. All the materials used for this were purchased at my local hardware store. The process is to make a physical pattern, make a mold from the pattern, and mold (actually it is a casting) the parts.

1. Create a physical pattern of the product. In this case, I carved a hand from a piece of ½" wood dowel. Then sanded and varnished it. You could also glue pieces together.

2. Using a clear plastic box like screws come in at the hardware store, mix up a batch of auto body dent filling putty and fill the plastic box. Set the pattern into the putty and let the putty harden. Don't forget to wax the pattern with a good automotive paste wax so the pattern will release from the putty.

3. When the putty cures, pull the pattern out and sand what will become the mold's mating surface (parting line) so this surface is flat. A belt sander is ideal for this job.

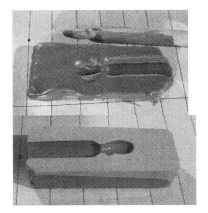

Prototype

The Practical Inventor

4. Place a release agent on that will become the mold mating surface. I used red packaging tape. The whole purpose is so the second mold half does not stick to the first mold half. Use a sharp razor knife to trim the tape.

5. Place the pattern back into the mold half.

6. Mix up another batch of the putty and fill another plastic box. Set the first mold half into the fresh mixed putty and let it set.

7. Before separating the halves, drill guide holes. It would work best to drill these on a mill or drill press so they are all parallel but if all you have is a hand drill, just do the best you can.

You can always clamp the mold together and press the guide pins in individually if that are not straight enough to slide the halves together. For the mold in the pictures, I used 1/8" plastic rod.

The Practical Inventor

Prototype

8. Carefully split the molds apart, clean up the two halves and wax the part cavity with a good quality automotive paste was.

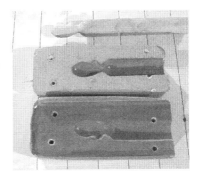

9. Install the guide pins.

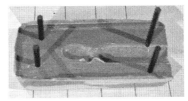

10. Set the mold halves together, clamp, and pour the plastic material in the mold. It works best if you have a plunger to force it into the mold and force out air bubbles. I used a piece of the wood dowel. Also, on this part, I used some vinyl tool dip but a two part material works much better for a reasonable cure time. Many plastics distributors carry a two part latex that works well.

11. Once the plastic is set, separate the mold and clean up the part as needed.

Prototype

The Practical Inventor

Building Prototypes

At some point, you are going to build some prototypes of your product. We have talked about many different aspects of building the prototypes. As a method of reviewing this chapter, I thought that it would be beneficial just to look at some examples of both prototypes and the process followed during the development.

The first example is an armrest design for a chair. First was a foam core mock up based on some pencil sketches and discussions with the client. This model was very useful because we were able to mount it on the chair and try it out before spending too much time and money. Also, if you look at the CAD model, you will notice the arm rest is wider and the cup holder has switched positions with the smokeless ashtray. We discovered a safety issue with reaching over the cigarettes and made the change. Had we not made the model and mounted it onto a chair first, we would have made a prototype and spent an extra $5000.

On this project, I created a 3D CAD model to test the idea and be sure everything would function properly. Since many of you don't have access to a 3D CAD system, you may opt to hire someone to model your part.

133

The Practical Inventor — Prototype

If you decide to do this, you should be sure to obtain the following in your contract:
- Written agreement with fixed cost, specific deliverable and time table.
- Deliverable is CAD model, fabrication drawings, IGES and STL files of the models.
- A CD with all your files.

Guarantee of support to fix any error that your designer might make. (This is called integrity and ethical business on your designer's part).

If your provider is not willing to provide al this in writing, find someone else. You may decide to purchase your own 3D CAD software. Just be advised, it is a pretty steep learning curve if you have no experience in this area. I suggest you visit a local rapid prototype fabricator. Most of them are glad to teach you a little about their processes, especially in hopes of obtaining some work from you.

The final step was to build an SLA model of the design. Most hobby stores carry really good, industrial strength super glue. Get both the thin and thick glue and the accelerant spray. Glue the pieces together and once the glue is set, sand the parts. (I suggest 400 grit wet/dry sandpaper) Paint or cover with fabric and the model is ready for show.

Another example is an antenna for testing some material processing results. While I cannot go into details of the project, this is a good example of a machined part as a prototype. Fortunately, in this case, it only had to work, not look good.

The final example is an assembly for another client where I had a quick turn circuit board, quick turn flex circuits and an SLA part. Once this was assembled to the level in this picture, this unit was assembled into a final assembly with several more cast parts

134

Exercise 6-3: Prototype Execution:

Well, this is it. In this chapter we have discussed the different method of making a prototype and the different uses for the prototype. We have also discussed contracting help and getting what you want for what you are paying.

The real question you need to answer now is whether you are going to build the next prototype or contract someone to build it for you. There is no black and white answer. Each of us has different experience levels and expertise. Chances are for most inventors, anything beyond a proof of concept prototype would be built by someone else. Since this is a chapter on prototypes, I would love to say it is time to build your next prototype but in reality, your next step is probably not bu9lding a prototype.

Go back and revisit the previous exercises in this chapter and then revise your development plan from chapter 5 on planning.

The Practical Inventor

Prototype

Chapter 7

Launch

Launch

This final chapter is not the end of the story; it is the beginning of life for your product and the world of possibilities opening up to you. If you've gotten this far, you should be applauding yourself. It is a huge accomplishment.

Let's pause for a moment and consider where you are. If you believe the product you are about to launch is going to make you a lot of money, then you must be thinking what you can do with that money. Don't spend it yet. Right now you still need to focus on getting the job finished.

This chapter is not going to teach you how to form a company, license a product or raise capital. We will address the basic items needed to create a scenario of when and where something should happen. A number of excellent resources are for helping you at each step of your journey are listed in Appendix A. One general repository of resources that has been exceptionally helpful is CEOSpace (www.CEOSpace.biz). This organization has helped me understand the capital (money) side of business, clarified business sequencing, and a great venue for finding world class professionals for my team.

At this point, you have established that you invention has a market and that the technology exists to produce it at a price that people will be willing to pay for it. You have also built a proof of concept prototype and secured your Intellectual Property but not filed patents yet. I talked to one inventor who not only has a prototype, a great video, and patents but also has production tools in process but doesn't have a team, market research or production plan. If you are reading this and saying to yourself, "Oh man, I'm way off track" don't worry about it. The major risk factor is not having enough capital to keep moving forward because you spent too much in the wrong place at the wrong time. Well, I cannot say what is right or wrong and sometimes the sequence will change giving specific circumstances.

What I am presenting is what I've learned over 30 years as to what works and what I've witnessed in successful companies. Remember the washed out trail? We just have to find another way around the washout.

The chart on the following page presents an outline for this chapter and a proposed sequence for you from this point forward. Some of these items may already be completed. If that is the case, review what has been completed or any opportunities to improve.

Launch   The
                                Practical Inventor

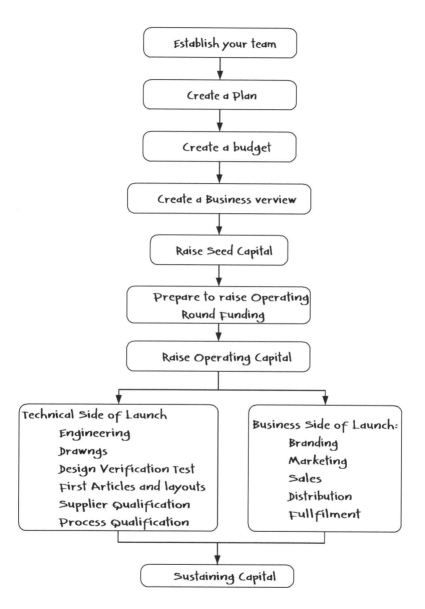

Let's look at each of these steps in a bit more detail:

Establish Your Team:

I like the Acronym for TEAM, Together Each Achieves More. This is so true! If it were not for my team, you would not be reading this manual right now. One of the rules for success is recognizing that we cannot do it all and need a team to help. Even if you have the ability to do it all, you likely do not have the time or desire. (I want to work on the things that I enjoy and find someone who really likes the accounting!)

Building your team will probably be the biggest sales job you've ever taken on. This is even more true if you think about the fact that you are going out to find friends (old and new) who will work for you for free to start. Here are some basic rules to start building your team:

1. Friends are fine but business is business. Put people into positions based on qualifications and performance, and then hold them accountable.
2. Write an agreement between each team member. The agreement should have at minimum:
    a. Their specific duties
    b. Your clear and achievable expectations for them.
    c. Timing expectations for achievement of duties.
    d. What they are going to get for their efforts. With ownership options be sure you don't give away the farm. With that, don't give each team member too much ownership and make this ownership conditional on your company reaching a certain point, such as shipping the first product. Also, double check with securities laws that you don't violate SEC laws by making his an official offering.
3. Each prospective team member should sign an NDA.

Remember, this is your idea and your company; you should always have the final decision. These people are working for you but listen to what they have to say.

Exercise 7-1: Building a Team

Refer back to Exercise 5-1 on pages 114 and 115, questions 25 through 33. Recompile the list of skills you will need for your project. Now find people your can rely on to fill in the skills you need. Remember, friends are nice but skills are better.

Launch   The
                                    Practical Inventor

Review and Revise Your Plan:

The plan we are looking at is not your business plan. You do not need that yet. Remember Chapter 5 where you put together a product development plan? This step by step process is what you need right now. Pull you plan from Chapter 5 out and keep it close while going through this chapter. Be sure to include your team and board on any changes.

> Remember, your plan is not chiseled in granite so don't be afraid of making minor course corrections in your plan.

Exercise 7-2: Planning

Get your team together and review with them your core values and the product development plan from Chapter 5. Make adjustments to the plan as appropriate from what you have learned since creating the plan and with input from your team.

Create a Budget:

Your budget is an ongoing living entity that guides you in distribution of funds. If you are working with investor money, you have a responsibility to control this distribution and to record it. A good budget does a couple things:
- o Provides guidelines for distribution of funds.
- o Provides a means of planning for capital requirements. The general rule for raising capital is this: Look at the budget and raise twice as much as you think you will need.

In terms of sequence, the general rule of thumb is that no one gets paid until the seed capital is raised. Then only the inventor, attorneys, and super-essential things are paid. Once you have competed the operating capital round paying for general services and other salaries is permissible in accordance with the budget.

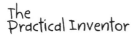

Exercise 7-3: Your first Budget

Refer to your product plan and assign an estimated cost to each step. Get you team involved helping build a budget. Start putting your budget together and keep working to refine it.

Create a Business Overview:

Many people think that you need a full business plan to raise capital. That is true later but not true for seed capital. A business overview is a only a few pages that explains:

1. What is the problem and your solution?
2. Who cares - what market?
3. What is the exit strategy - potential target for ROI 10x - 20x return?
4. Can you make product X?
5. Can you sell or market this?
6. Can you manage and produce this?

Many people who have been in the business of raising capital for many years will look for a 10-16 page document as a business overview. Appendix D has a model for such a business overview that I would suggest that you start working toward.

Exercise 7-4: Your first Business Overview

Get your team together and answer each of the 6 questions:
1. What is the problem and your solution?
2. Who cares - what market?
3. What is the exit strategy - potential target for ROI 10x - 20x return?
4. Can you make product X?
5. Can you sell or market this?
6. Can you manage and produce this?

Launch   The
                                    Practical Inventor

Raise Seed Capital

This is not funding to build your next prototype. This is the capital you need to pay yourself and your attorneys to get things in place to raise your operating capital which will be used for finish development and launch of your invention. What many people do not realize is that it takes capital to raise capital. The seed capital is what you use to raise the operating capital. Teaching you how to do this is beyond the scope of this manual. There are resources in Appendix A that specializes in raising capital. I would also encourage you to shop around before committing to any one person. A really big part of it is for your personality types to match. If you are not comfortable working together, it will not be good business for either of you.

Prepare to raise Operating Round Funding:

Now you will need a lot of pieces because you will be raising a million dollars or more. You will need to have a full business plan based on your business overview, securities compliance, and many other pieces that right now your head is probably spinning and you are asking yourself what you are getting in to. Think of the book; Hitchhikers Guide to the Galaxy, open the front cover and it says in big letters "DON'T PANIC". This is why you have a team, use the resources that are available.

Raise Operating Capital:

Part of the reason for raising the seed funding and paying yourself is that this step is a full time job if you are serious about your project. At this point, you have to make a major decision: either do it or don't do it. Absolutely do not just play with it part-time. If you have gotten this far, your invention is very likely a good product. Understand that most good products do not launch successfully for one reason LACK OF CAPITAL. Don't be one of those.

Launching:

Once you have raised your Operating Capital, implement your plan in accordance with your budget. How would you feel knowing that you are now unstoppable and able to implement one of your dreams? (If you want an idea of how it makes me feel, just think D10 Cat.)

This stage I have divided into two segments in general skill sets. One is technical and the other is business. This is where you could separate between a CEO (Chief Executive Officer), COO (Chief Operating Officer), and CTO (Chief Technical Officer) to move your project forward faster and more effectively. At this point, many inventors opt to move into the CEO/CTO role and bring in someone for the COO side. The different segments have the following different functions:

- **Technical Segment**
    - Engineering
    - Drawings
    - Design Verification testing
    - First Articles and layouts
    - Supplier Qualification
    - Process Qualification
    - Manufacturing

- **Business Segment**
    - Branding
    - Marketing
    - Sales
    - Distribution
    - Fulfillment

While many people will want to divide up these segments slightly differently, I think you get the concept.

Exercise 7-5: Organize Thyself

1. Take your product plan and rearrange it to reflect what we have discussed in this chapter about the sequencing.
2. Have each member of your team re-evaluate their involvement and placement in the plan.

Sustaining Capital:

My business mentor of many years, George W. Harding, was an amazing business coach. He had been on the Board of Directors of General Motors; Director of the War Production Board for non-ferrous metals in World War II, at one time owned Vickers Hydraulics and Chris Craft Boats amongst many other adventures. He would constantly remind me of 2 things. First that money is only one tool in your tool box and that is all that it is. The second thing was that the pulse of a healthy company is cash availability. If you have extra cash, you may be in trouble.

A healthy, growing company almost always needs more cash than it has. Why is this? Simple, if you have $100,000 in receivables, $100,000 on store shelves, $100,000 in warehouse, and $100,000 in process, that is $400,000 tied up. What happens when you take on a new account with an additional $150,000 sales, you now need another $600,000 in operating capital to support these sales. By the time you get here, your CFO has built your Balance Sheet so you can obtain a line of credit to support your sales. This is a very typical and normal situation so don't stress over it. You just need to know about it now so that once you get here, you can send me an email and thank me for warning you early. Then we will get together but forget the coffee, let's have a beer and some Calamari.

Licensing VS: Manufacturing:

We really have not talked much about licensing or manufacturing but have focused on building a business around your invention. You may be able to license your product at any stage in the development that we have been discussing. This can be a tricky business. If you go to market and demonstrate with sales, the validity of your invention, while you are developing value of your intellectual property and your invention, if you go to market and demonstrate with sales the validity of your invention, this can strengthen or weaken you ability to license your invention in the most favorable manner. I strongly suggest that you engage the services of a licensing expert within the field of your invention. Just be very careful because you could be left with nothing very easily. You should not be paranoid but at the same time, don't be naive either.

Here are some basic rules for licensing:
1. The up front license fee is often close to the development cost.
2. The royalty structure varies. I always shoot for being paid a specific amount for every item that ships. Not on gross sales and definitely not on net profits.
3. Do not discount your royalty for volume. Many licensees will try to have you reduce your royalties as sales increase. Why should you penalize yourself for being successful?
4. If you give an exclusive license, be sure there is a performance requirement with your right to retract the license.
5. Sometimes the first deal is not the best deal.
6. Rarely is the deal perfect.
7. The license is your permission to allow someone else to use your intellectual property. Don't give away the farm but also don't expect a check for $10 Million either.

Listen to your license agent as to what is normal and customary in that industry, if you win, they win too.

Launch   The
                                          Practical Inventor

Happy Trails

OK then, we are at the end of the manual but hopefully not the end of your journey. Please revisit the different exercises as your project progresses. Remember, developing your product is a dynamic entity and subject to change, not only due to your understanding but also market conditions.

Our sincere hope and desire is that this manual and The Practical Inventor kit have been helpful to you. However, if you have found things that could be improved upon or would have helped you along the way, please share these with us so we can better help others. You can contact us anytime through our website www.thepracticalinventor.com.

Happy Inventing

The
Practical Inventor   Launch

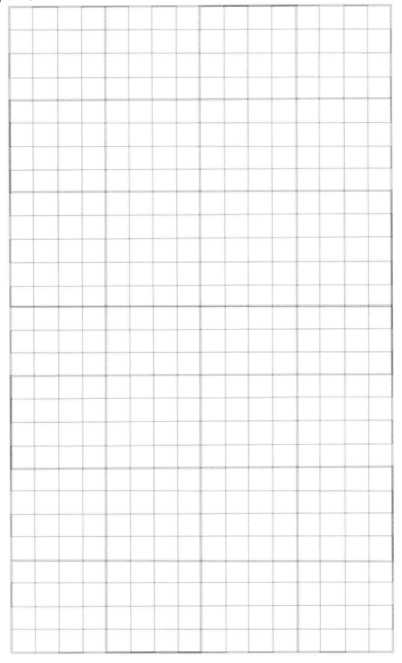

Appendices

Resources
Inventing the Inventor
Business overview Model

Resources

Management Consulting

CEO Space – CEO retreats and training, networking, business development
www.ceospace.biz
www.ibiglobal.com

SBDC – Small Business Development Centers
www.sba.gov/sbdc/

SCORE - Counselors to America's Small Business
www.score.org

Hoyt Management Group - Strategic planning and business development
100 Ferne Ave.
Palo Alto, CA 94306
(877) 367-4698
www.hoytgroup.com

Advanced Management Planning & Project Mapping – Project planning
Randolf Craft
Pacific Planning Institute, Inc.
93B Puako Beach Drive
Kamuela, HI 96743
www.planinparadise.com
planinparadise@aol.com

The Practical Inventor

Engineering

Veritek, LLC – New product development, engineering, consulting
PO Box 630253
Littleton, CO 80163
jtw@veritek.com
www.veritek.com

Manufacturing

Veritek, LLC – Contract manufacturing, supply chain management, new product launch, sheltered workshop for Disabled American Veterans
PO Box 630253
Littleton, CO 80163
jtw@veritek.com
www.veritek.com

Prototype Design and Construction

Veritek, LLC – prototype design, construction, transition to manufacturing
PO Box 630253
Littleton, CO 80163
jtw@veritek.com
www.veritek.com

Rapid Prototype Components Fabrication

Protogenic, Inc. – rapid prototype component fabricator
1490 W. 121 Ave.
Suite 101
Westminister, CO 80234
(303) 453-3990
bob@protogenic.com
www.protogenic.com

The
Practical Inventor   Resource

Legal
Brian Kunzler – Intellectual Property, Patents, Trademarks
Kunzler and McKenzie
8 East Broadway
Suite 600
Salt Lake City, UT 84111
(801) 994-4646
Kunzler@kmiplaw.com

Maria Speth – Trademarks, intellectual Property litigation
Jaburg & Wilk, PC
3200 N. Central Ave.
Suite 2000
Phoenix, AZ 85012
(602) 248-1089
mcs@jaburgwilk.com

Karl Dakin – Licensing, contracts, securities
Centennial, CO 80016
(303) 916-8272
Info@KarlDakin.com

Note: This is not an endorsement of any of these service providers and should not stop you from your due diligence.

Appendix B
Inventing the Inventor
Co-written JT Wilkinson and JM Cross

Do you have dreams of things you want to do, who you want to be? Are you living your dream? If you are not living your dream there are likely a lot of excuses and a few reasons. The biggest reason most of us are not living our dreams is that it takes a great deal of courage and tenacity to step out. Chasing your dream is uncomfortable and can be unsettling when you are stepping out of what feels a safe and secure position. Say that you always wanted to have your own fly fishing store but it is hard to leave that cushy $20 per hour factory job, the regular pay checks, the insurance. It feels more secure than going it on your own. It is more a perception than reality so we tend to make it reality and fear stepping out. Should you quit your job, sell everything, and follow Obi-Wan Kenobie on some foolish crusade? I don't think so, in fact, don't quit your day job unless you are independently wealthy and just work for something to do. If that is the case, my contact information is in the appendix, we should have coffee.

Exercise B-1:

This has got to be one of the hardest, yet most fun exercises that I am going to ask you to do. Also, don't expect to have this exercise completed; in fact it is likely to become a life process instead of just an exercise. Get out a pad of paper or preferably your design journal.

1. Acknowledge your dreams – make a list of your dreams. Don't limit yourself, if it is your dream, write it down. First manufacturing engineer on the Space Station is still on my list. Read the book "The Power of Intention" by Wayne Dyer.

2. Write out each of your dreams in as much detail as possible. I encourage you to add sketches and pictures that help define and materialize your dream. Cut and paste photos from magazines to help you visualize the dream.

3. Take a first step. The difference between a dream and a goal is action. Within each dream, write down what steps you can take and when you can take them. Don't worry about it if you cannot see the entire path to your dream. It is kind of like taking a hike on a mountain trail. Often you can see your destination but not each step so all you can do is take one step at a time and as you come over the next rise, deal with what is in front of you at the moment. Just keep taking steps forward. Don't get discouraged if you have to backtrack around a rock slide or washed out trail. Life happens.

4. Don't resist the barriers that are keeping you from living your dreams. Visualize yourself in the place of your dream. Remember the washed out trail? Just go around it and keep on moving forward.

5. Find the people who will support you. Someone told me that we are the sum total of the 5 people we spend the most time with. We all have the people who are so pleased to pop your dream and criticize you. Just choose to spend time with the people who will support you, not drag you back into the pit with them because they are too afraid to climb out of the pit for themselves.

6. Applaud your successes. Reward yourself, even for the little victories. The biggest victory of all is that first step forward.

Be a student of success. One of my favorite books is Think and Grow Rich by Napoleon Hill. If you want to see how some of the most successful people of the early 20th century overcame adversity, kept their eyes on their dreams, and pushed through, you will enjoy this book too. Pick your most favored and admired successful people and study them, read their biographies.

Personal Introspections:

Over the years I have come to realize that all things on earth, whether it is a human being, idea, business or product – has its own "life". The "life" of your invention – just like your own life has a "body, mind, and spirit". The "body" is the end product or service you are delivering to the customer. The "mind" is the "business" aspects of your product – sales, marketing, production, etc. And finally, the "spirit" is the individual personal capacity of the inventor. "Inventing the Inventor" focuses on the "spirit" of your invention – the Successful Inventor.

It's been said that you can't have a successful product or business without first being a successful person. Now this doesn't mean that you already need to have fame or fortune to be a successful inventor – it means that you need to have developed the personal capacity to carry your dream from idea to marketplace.

Dictionary.com defines an "inventor" as someone who invents (Duh) or a person who produces or contrives something previously unknown by the use of ingenuity or imagination. As the mother of seven children (five of whom are boys) *my* personal theory is that we all are born inventors – especially in times of boredom or desperation!
The difference is that a *Successful Inventor* creates something that has mass appeal or use and hopefully brings the Inventor the personal satisfaction, acclaim and a lot of money as a result. So let's look at some of the personal qualities it takes to be a "Successful Inventor".

I = **Intention** – the fuel that takes your product from idea to reality

N = **Now** – staying present – yesterday's over and tomorrow never comes

V = **Versatile** - ability to adapt and change tasks easily.

E = **Excited** – your excitement is contagious and drives the idea

N = **Networked** – connected to community at multiple levels

T = **Thorough** – willing to do the work it takes without shortcuts

O = **Optimistic** – having a positive outlook and confident about your idea

R = **Resilient** – your ability to rebound after disappointment

The following sections discuss each attribute and guides you through the process of seeing your own strengths and identifying areas of your life to improve and build your personal capacity to make your invention a reality.

Intention

All of us have "invented" things – a new way to put on our socks with a coat hanger when our hand is in a cast or a new way to get our work done when a deadline is looming. And we all know the saying – *desperation is the Mother of Invention* and watched people do some pretty "inventive" things – both good and bad – to solve a problem. The difference between all the little "inventions" in everyday life and being a Successful Inventor is intent – our commitment to do what it takes to create a product or service that will have maximum impact.

Intent however is more than just a decision – it encompasses vision, personal discipline, integrity, and accountability.

Vision

Inventing the Inventor — The Practical Inventor

My youngest son, who is now 11, recently read the story of the Wright Brothers for a school book report. I had forgotten, as I think most of us do, that the "Flying Machine" (airplane) was just the most famous of the Wright Brothers inventions. The story goes that their first "invention" had to do with the production of a sled with a rudder that was faster and more maneuverable than any of the other boy's sled in town. It was in that moment of "flying" down the snow covered hills that the vision of flight was born. What exactly is vision though?

Recent studies in the application of quantum physics principals indicates that "vision" can be thought of as a field – similar to gravity. Your *vision or purpose* will attract the information, people, and resources you need to become more than an idea. So…what does this all have to do with your invention? Simple, the more clearly defined your vision is, the more effective magnet it will be in attracting what you need to make your invention a reality in the marketplace.

When I coach people the first question I ask them is what do you want? This question most often leads to my biggest pet peeve – their answer. First they stammer a bit, usually from the shock of someone actually asking them the question which is compounded by the fact few people ever take the time to really define what exactly they do want in life. Most of the time they think its more money or things, however when you quiz them a while even that vanishes in a sea of hems and haws. So ask yourself this question – what do I **_REALLY_** want in my life?

One of the best exercises I have found to clearly define your vision and carry it around with you in various forms to help you maintain your focus and intention is what's called the Perfect Day.

THE PERFECT DAY EXERCISE WILL HELP YOU DEFINE AND CARRY YOUR VISION THROUGH THE PROCESS OF MAKING YOUR IDEA REAL!

Step 1: Tell the Story - It's important that you first answer the questions below verbally into a tape recorder or even into your telephone answering machine. The objective is to be able to listen to it over and over again to help you carry the vision throughout the process of achieving your goals. I play the "tape" when I'm overly frustrated or tired to remind myself where I'm going and what it's going to look like when I get there. It always rejuvenates and jump starts my motivation.

Step 2: Create the Picture - After you have recorded the answers (your vision of a Perfect Day) then get yourself a poster board and a stack of magazines, catalogs, etc. and create a picture (visual representation) of your

Perfect Day. Once your poster is complete, take a digital picture of it – you can then print it in various sizes and post it anywhere and everywhere you spend time as a constant reminder of what you are working for. I have a copy on my bathroom mirror, on the refrigerator and one in my calendar/appointment book. This gives me the opportunity to get a quick visual reminder of what I'm working for daily. Again, when things are bad it reminds me of where I'm going and it eases the struggle. When things are good it helps me recognize how close I am to realizing my dream and living my Perfect Day everyday.

Step 3: Write it Out – There's an old adage that says if its not on paper its not real. When it comes to your vision, putting it on paper does two things. First, the kinesthetic process of putting your thoughts on paper strengthens the commitment in your mind. It also allows you to share the vision with others and can serve as the foundation for your business plan. I have made it a habit to re-write "my perfect day" in my personal journal at least once a month to cement the vision in my consciousness.

Personal Discipline

As simple as it may sound one of the biggest things that "stops" inventors from succeeding is the lack the personal discipline. I have met more than one person with great ideas who never move beyond the dream stage simply because they lack the personal discipline to make it to an appointment on time or to follow through on a lead. Or they get side-tracked with things that bring immediate gratification without ever spending time on what will assure long-term abundance and satisfaction.

There are a hundred tools and systems to help you manage your time – bottom line though is that you are the only one who can make you do anything – that's why they call it *personal* discipline! One of the easiest things to do to strengthen your personal discipline is an exercise created by Thomas J. Leonard called "The 10 Daily Habits". In short, the point of the exercise is to help you both become more conscious of the things that you need in your life to have a positive, productive day and also to help you build the consistency needed to integrate these activities into your daily routine.

Step 1: Take a week and begin to jot down those things you do on days that you are especially positive and productive. They don't need to be especially technical or big things. Just little things you can do each day to make things better for yourself and keep yourself on track. For example, some of things that help me have a much better day include taking my vitamins each morning with a smoothie; taking a walk in the late afternoon; a bubble bath at the end of the day. Like I said – the simple little things that make your day better.

Step 2: After taking a week or so to really NOTICE what helps your day go better – **make a list of ten** *of those things. Remember, it's not about huge tasks or extra hard steps, just ten little things you can do for yourself to help your day be more positive and productive. Once you have your list then its time to "play the game". The objective is to get 70 points a week – one point for each of the ten items on your list for each day of the week (7x10 = 70). Play the game and chart your progress for three months (90 days).*

Research indicates that if you engage in a specific behavior – good or bad – consistently for a period of nine to twelve weeks you will establish a "habit". This is why so many "programs" whether its boot camp or AA use the 90 day framework. Creating positive habits is a way to continue to strengthen your capacity to do what it takes to make your dream come true. (Just a side note – when I notice my life getting a bit off track I go back and "play the 10 daily habits" game to get me back on the right path. It's a nice tool to help you refocus when things get crazy!)

Integrity

Integrity is simply "walking the talk". Do you say one thing and then do another? Are you someone that can be trusted to say what you mean and mean what you say? Like personal discipline, your integrity is something only you can control. Integrity begins with knowing who you are and what you stand for – your values, standards, and ability to be authentic.

Many of us have heard the old saying "if you don't know what you stand for you'll fall for anything". Knowing yourself and what you stand for is the first step in developing a personal code of ethics that provides the armor you will need when you are assaulted by the "dragons". Dragons are what I call the array of ethical challenges inventors face on a daily basis. Most of the issues you will face fall into the following categories:

- o <u>Fidelity</u> – Did you make a promise/contract (implied or expressed) that you failed to make good or need to uphold to stay in integrity?

- o <u>Reparation</u> – Have you committed a wrong (even a little one) that you now have to make up (even if you're the only one who knows about it)?

- o <u>Gratitude</u> – Are you grateful for something someone did for you and failed to appropriately thank or reward them?

- o <u>Justice</u> – When making choices that effect others are you willing to take an objective stand for fairness without bias?

- Benevolence – When you have the ability and opportunity to help someone out who needs your help, do you?
- Do no Harm – Do you take all care to avoid harming anyone (yourself included) unnecessarily through your words, actions and decisions?

There are an array of ethical standards and codes of conduct for professionals of all types. My recommendation is that you review the websites of the appropriate industry or profession related to your invention. This is a good place to start however it is also important that you infuse these codes of ethics with your own standards to make them yours.

This "code" will be a welcomed shield when in the throes of battle with the "dragons".

Accountability

The final ingredient in your intention is accountability. Being a Successful Inventor means being willing to be accountable for your words and deeds – to do what you say you will do, when you say you will do it and be responsible for and deal with the consequences of your decisions and actions good or bad. It's easy to take credit for the positive things in our lives. It's another thing to stand up and be counted when things didn't go as planned. There are times we fail. Successful people learn from these failures and do not let the failure identify with them. Being accountable is learning to face the consequences of our actions. After that, learn from them, hitch up your britches, and get on with it.

Exercise B-2

THE FOLLOWING ARE THE QUESTIONS TO HELP YOU DEFINE YOUR PERFECT DAY – THIS IS NOT A COMPLETE LIST, IT IS JUST A FEW QUESTIONS TO HELP YOU GET STARTED IN CREATING THE MOST DETAILED DEFINITION OF YOUR VISION OF THE FUTURE POSSIBLE.

1. How does your perfect day begin? What does the world around you look like – are you living in the same place you are living now or somewhere else? Who is there with you – family, friends, people who work with you or for you, etc.? What does your home and/or office look like? Fill in as much detail as you can to set the scene for your perfect day.

2. What kind of things do you do during your perfect day? How much time you spend working, with family and/or friends, being involved in the community, in worship, etc.? What are the activities like – challenging, fun, fulfilling, etc.?

3. What are the outcomes of your perfect day? How many customers do you have? How much do they buy? What do you use the money you are earning for – those things you always wanted to buy, support and gifts for your family and friends, philanthropic activities, etc.?

4. What are all the activities or steps you've taken to realize your perfect day? How long did it take you to design, develop and launch your invention? Who are all the people who helped you reach your goals? What was the thing – i.e. the first check or big contract, getting your product on Home Shopping Network, etc. – that happened that took you "over the top" or was the one thing that made you realize success was here now?

Now

To be a Successful Inventor you have to be willing to do it NOW! Ideas tend to float around until they find someone willing to do it NOW – to take the idea and turn it into a marketable product for the masses.
We all have heard someone saying "I thought of that *year's* ago" when a new product hits the market – the difference between that person and the guy who put it on the market is that he did it NOW!

What's keeping you from working on your invention NOW! What are you waiting for? Working on your invention doesn't mean you've finished it – it means taking decisive action each and every day to move you forward. Whether you have capital or not there are things you can do – design sketches, market research, or simply taking a walk and visualizing the next step for your project. The point is to be IN ACTION NOW!

Versatile

Being a Successful Inventor means being able to turn easily from one task to another, especially if you are a solo-inventor without a huge staff of people to help you get your product to market. This doesn't mean that you have to *do* everything – it just means that not only are you the one who thinks up the idea but you are also the one who cleans up the shop at the end of the day until you sell the idea and can hire someone.

There are some great tools and strategies to help you manage the various things you have to get done each week. One of the best is I've found is to group activities according to geographic and content. For example, I block out every Monday just to do the paperwork, Tuesday's for meetings with my staff, Wednesdays for meetings with clients, etc. I then schedule things that come my way to fit in a slot on the appropriate day. By the way – I also schedule Friday afternoons just to have fun, a key ingredient in keeping you healthy, motivated and creative!

Excitement

Your excitement is the fuel for your vision. It is what engages others in your work and keeps you motivated daily.
 This could also be referred to as your passion for pursuing your dream.

Networked

Be connected to community at various levels. There is an old adage that says "It's not what you know, it's who you know." There is a lot of truth to that. As you start to build your team, having a network will be an invaluable resource. Who knows, you might even be able to help others in your network as well.

Thorough

Willing to do the work it takes to see your dream to reality. It seams that the closer you get to completion, the harder it is to make process. Being successful requires that attention to detail and following through step by step to completion.

Don't take shortcuts. You will get a lot of satisfaction and build a strong foundation by doing the right things and doing things right.

Optimistic

Have a positive outlook on life and your idea. This is an attitude. You decide every morning that today will be a great day.

Resilient

What is your ability to rebound after disappointment? Success is being willing to accept there are roadblocks and sometimes you just have to take a different path.

Professional Introspections:

As a successful business person, there comes a time when you must come to terms with the harsh reality that you cannot do it all even though you might really be able to do it all, there are just not enough hours in the day. I know a lot of inventors who feel that not being able to do it all is an attack on their character. I used to feel that way too. The reality is that the most successful leaders know how to recognize their weaknesses and the things that they could delegate (outsource) to someone else. Ronald Reagan is known as the greatest president of our time when he did very little. What he did however, was to surround him with functional experts, empower them, and get out of their way. Follow President Reagan's example. Put on your team functional experts, empower them and get out of their way. I finally learned that and can hardly wait until I finish this manual so I can get back into my lab. Since figuring this out, it has taken almost a year and a half but my business is getting ready to bust the doors down. What about you? What are you willing to let go of? How much is that letting go and being able to focus on the things you love doing worth?

Exercise B-3:

1. Make a list of the things you do to move your invention forward on a regular basis:

2. Now cross out the things you love doing and are very good at. What is left are the activities you need to find someone to take over, give them clear goals and instructions, peek in on them from time to time, and get out of their way. You will be amazed at the cool things they will do. Unfortunately, some of them will perform poorly. You will have to replace them. DO NOT keep them in that position because they are friends.

We wish to leave you with 2 thoughts:

1. The axiom "no man is an Island until themselves" is very true. Get help where you need it. Recognize this as a position of strength, not weakness.

2. Teamwork makes the Dream work. Put together a high performing team and move to a position of success with your invention.

Appendix C

Business Overview

By John Hessenbruch

A business overview is actually an abbreviated business plan. Typically, the length of the document is between 10-16 pages. In addition to providing a market evaluation of the product, the overview summarizes the business opportunity for a potential investor. In order to make the document more "user friendly" it is highly recommended that key figures, tables, and charts also be included as support documentation.

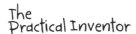

Most business overviews include the following information:

- Abbreviated Executive Summary (1-2 pages)
- Technology Description
- Target Industry Overview and Potential Customers
- Market Summary and Product Development Strategy
- Competitive Analysis
- Operating, Manufacturing, and Sales Plan
- Management Team
- Financial Summary

Abbreviated Executive Summary

The purpose of the Abbreviated Executive Summary is to immediately capture the reader's attention regarding the proposed opportunity. Generally, a paragraph should be included that describes each section of the Business Overview. In short, state the problem, proposed solution, what's in it for them, and the exit strategy.

Technology Description

A concise Technology Description should be written that provides the investor with a clear overview of how the new product can either improve or revolutionize the current technology in the marketplace today. A brief overview should be included regarding the new product's competitive advantage to other products on the market and how the new product will fill a strategic need within the target industry(s). It is suggested that one or more key figures be used to describe the innovation.

Target Industry Overview and Potential Customers

An overview of the target industry(s) and potential customers should be provided that addresses the following questions:
- What segments of the industry will use your product and why.
- How is each segment (or industry) structured?
- What are the primary issues or problems that are apparent in each industry segment and how will your product provide a solution with respect to: Cost Savings, Speed, and Efficiency?
- Which companies are the major firms in each industry segment?
- Which companies within each industry sector would be considered as potential customers?

Business Overview   The Practical Inventor

- Which companies have you met with and how will your product help those firms attract more business or make them more productive?

Market Summary and Product Development Strategy

What is the size of the market of each target industry that the Company intends to distribute its new products? In particular, the overview should describe how the target industry has expanded over the last 5 years and how the industry will grow over the next 5 years. Pie charts or bar graphs can be used to show how each target industry (or industry segment) is subdivided with respect to revenues and growth potential. They can also be used to show which industry applications (commercial or military) may be more important for the Company's future growth.

The Product Development Strategy should be addressed with respect to product mix, product life cycle management, and (if possible) brand name/brand image. In particular, a discussion should be included regarding the marketing strategy with respect to the new product, the potential customers, and (if applicable) geographical market.

Competitive Analysis

A Competitive Analysis must be generated to provide the investor with an overview of the primary firms that dominate each industry (or industry segment). A summary must be included that discusses the advantages of the Company's new product with respect to the competition regarding cost, performance, reliability, service, and applications. In addition, a brief evaluation of how these advantages can be leveraged by the Company in the future.

Operating, Manufacturing, and Sales Plan

A summary should be generated that outlines the proposed operating plan over the next 3-5 years. The overview should include whether the Company intends to outsource or provide its own in-house manufacturing. A brief overview should be added that describes the sales strategy regarding promotion, sales force requirements, publicity, and advertising. If applicable, a discussion should also be included on the sales distribution requirements with respect to supply chain management, logistics, and channel partners.

The Practical Inventor

Business Overview

Management Team

A brief overview of the current management team should be addressed including a summary of all directors. In addition, an overview should be provided when the key management personnel will be hired (i.e., President, VP of Sales, VP of Marketing, VP of Operations, VP of Manufacturing, etc.)

Financial Plan

Projected financial statements provide an overview of a business' profitability and financial condition. in order to give a potential investor an accurate represention of the proposed venture, it is recommended that a mininmm of three schedules be prepared over a 3-5 year period.
Cash Flow Statement (Revenues-Expenses): summarizes how and when the investor can be reimbursed.
Income Statement (Profit/Loss Schedule): outlines a company's operational results on a monthly and annual basis.
Balance Sheet (Asset/Liability Statement): provides an annual overview of a company's assets, liabilities and net equity.

For the Business Overview, the Cash Flow Statement is probaby the most important schedule because it summarizes how much capital is required for the initial investment. It also demonstrates the return on investment (ROI) and the internal rate of return (IRR) for the project after a 3,4,or 5 year period. ROI is a key "yardstick" used by most investors to determine the compounded annual rate of return their investment will generate over a certain number of years. The IRR is computed on the initial investment and demonstrates the annual return on investment based on the cash flow over the 3-5 year period

For a more detailed description along with step-by-step instructions for creating your business overview, look for The Practical Entrepreneur's Business Planning Manual ISBN: 978-0-9817954-1-6 in your local bookstore or at www.thepracticalinventor.com.

The Practical Inventor

Other products by Entrepreneur Resource

If you've enjoyed **The Practical Inventor** and found it useful, you may also find our other publications helpful:

The Practical Entrepreneur's Guide to Business Operations

All profitable companies have business processes to execute business operations for delivering your product or service to the customer. The challenge comes in knowing what these operations are.

The Guide to Business Operations guides the entrepreneur step by step to:
- Discover and document existing processes
- Understand what proper and effective process documentation is.
- Develop a plan to develop a comprehensive documented operations management system based on the ISO-9001 International Standards
- Implement your plan.
- Integrate Six Sigma and Macomb Baldridge Quality Management principles into your system.

The Practical Entrepreneur's Guide to Business Planning

Every business starts with an idea. This new business has a much better chance of success when there is a good plan for launching and developing the business around the idea. Our Guide to Business Planning guides you step by step through the planning process. The planning process flows through:
- Executing Overview – one page
- Business Overview – 10 to 16 pages
- Business Plan – 20+ pages

Learn what each of these is used for, when they are appropriate and how to grow your plan from step to step then let us help you.

The Practical Inventor
The Practical Entrepreneur's Guide to Human Resources

Eventually, you will want to start hiring employees. Human Resources are every bit important as you capital resources. Walk through our step-by-step process of developing the resources appropriate and necessary for your company including:
- Employee policies and policy manual.
- Interviewing potential employees.
- Hiring best practices.
- Discipline procedures.
- Understanding and complying with the EEOC requirements.
- Navigating the health insurance.
- Determining appropriate pay and benefits.

Check our website at www.Veritek.com for the latest information.

The Practical Inventor

The attached Practical Inventor note is good for $50 off the cost of a Practical Inventor Kit only when purchased directly by mail from Entrepreneur Resource, LLC. To redeem, remove this page and return as the order form. This page must be included, duplicate copies are not accepted.

Send your payment of $199.95 US + shipping to Entrepreneur Resource, LLC, PO Box 630253, Littleton, CO, 80163 USA

Name _____

Address _____

City _____ State/Province _____

Postal Code _____

Email _____

Shipping: $24.95 US shipping, check website for international shipping.

Additional shipping notes:

_____
_____
_____
_____
_____
_____
_____

The Practical Inventor

Made in the USA
Charleston, SC
18 May 2013